Urban Peregrines

Urban Peregrines

Ed Drewitt

Pelagic Publishing
www.pelagicpublishing.com

Urban Peregrines

Published by Pelagic Publishing
www.pelagicpublishing.com
PO Box 725, Exeter EX1 9QU

ISBN 978-1-907807-80-0 (Pbk)
ISBN 978-1-907807-81-7 (Hbk)
ISBN 978-1-907807-82-4 (ePub)
ISBN 978-1-907807-83-1 (Mobi)

Copyright © 2014 Ed Drewitt

This book should be quoted as:
Drewitt, E. (2014) *Urban Peregrines*. Exeter: Pelagic Publishing.

All rights reserved. No part of this document may be produced, stored in a retrieval system, or transmitted in any form or by any means, electronic, mechanical, photocopying, recording or otherwise without prior permission from the publisher. While every effort has been made in the preparation of this book to ensure the accuracy of the information presented, the information contained in this book is sold without warranty, either express or implied. Neither the author, nor Pelagic Publishing, its agents and distributors will be held liable for any damage or loss caused or alleged to be caused directly or indirectly by this book.

British Library Cataloguing in Publication Data
A catalogue record for this book is available from the British Library.

Cover image by Sam Hobson (samhobson.co.uk)

Contents

About the author vii
Foreword ix
Preface xi

1 The Peregrine 1
2 What is an Urban Peregrine? 33
3 How to Spot a Peregrine 41
4 A Year in the Life of an Urban Peregrine 48
5 Food and Feeding 75
6 How to Study Peregrines 104
7 Ringing Urban Peregrines 125
8 Myths about Peregrines 140
9 Changing Threats and the Future of the Urban Peregrine 145
10 People and Peregrines 162
11 Where Next? 187

Further reading 193
Acknowledgements 195
Photographers 196
Index 203

About the author

ED DREWITT IS A PROFESSIONAL NATURALIST, wildlife detective, learning consultant/trainer, and broadcaster. He has been studying urban Peregrines for over 15 years, specialising in colour ringing their chicks and identifying what they have been eating.

Ed spends a lot of his time showing people wildlife, specialising in teaching birdsong, and helping others to identify, appreciate, and get hands on with nature. He also takes people around the world on holiday tours to see a variety of animals including whales, dolphins, and a variety of birds.

Ed Drewitt (Tracey Rich)

Foreword

The fastest bird on earth – more than ten thousand bird species but none moves quicker than this. And it's beautiful – in a predatory and purposeful way, in a way that screams power. Broad shoulders, sharp wings, massive talons, big eyes and a killer beak; this bird means business. For thousands of years we have revered, feared, loved, hated, reared, worshipped and killed it; it's never been overlooked or ignored, it's a part of our culture and enjoys a sensational synergy of respect and awe. From the tombs of pharaohs to the emperors of China, the princes of Europe and now to the precincts of urban Britain the Peregrine Falcon is the impressor. I don't know why superheroes ever bother to get out of bed.

There is one cachet it has forfeited to join us in the 21st century manscape, and we're thankful for it: its rarity has faded, and in many ways so has its inaccessibility. And that should secure its future as thousands now watch and admire the intricacies of its behaviour and ecology on remote cameras. On the rainy streets of Norwich, the sunny sidewalks of Chichester and the bustling pavements of Bristol, we can stop shopping and watch the best avian royalty reality show on earth. As Ed says . . . 'Wow!'

And what Ed says is worth reading. His obsession with the species and enthusiasm to know it better fuels this rich and comprehensive look at the Peregrine and its place in the urban landscape. The diversity of material is marvellous: history, habits and full-on scientific analysis, opinion, even postulation and essential practical advice . . . it's a feast of all things Peregriney, perfect for the young observer who's just seen their first falcon or the expert who is guaranteed to learn something new. The photos are good too.

So read it and get to know your most exciting new neighbours in the outdoor penthouse suite, forgive their table habits and keep your eyes peeled for a glimpse of speed as a bird splits the sky and stirs your heart.

Chris Packham, New Forest 2014

Preface

'Whoosh', 'wow', 'incredible' – just some of the reactions as a Peregrine powers over an audience at the Hawk Conservancy Trust in Andover, Hampshire. With eyes on the lure, the falcon turns swiftly in the air and at full pelt flaps its wings at a blurring speed to grab its food. Just as the Peregrine is about to take it, the lure is whipped away by the falconer, and the falcon repeats its manoeuvres. Finally, after some important exercise, the Peregrine is allowed to catch the lure and bring it to the ground. It mantles the food to stop anything else stealing its well-earned prey and gulps it down in oversized mouthfuls. This is just one memory of many I have of watching Peregrines when I was younger, at a time when seeing one in the wild where I lived in Surrey was unheard of.

Peregrines, like dinosaurs, big cats and owls, provide a great lure and gateway for getting lots of people interested in nature. They catch people's attention, wow audiences with their ability to kill and to fly at fast speeds, and inspire people to get more involved with the natural world.

In 1991, when I was 11 years old, I visited Symonds Yat Rock in the Forest of Dean, Gloucestershire, to see my first wild Peregrines. Wow! Real life Peregrines – I remember reaching up to look through the scope, and seeing at least one bird at the nest site. That week I had also visited The International Centre for Birds of Prey (formerly the National Birds of Prey Centre) in Newent. Jemima Parry-Jones was leading the display there – I was right at the front cross-legged on the grass watching the raptors being flown, including a Bateleur Eagle and a Secretary Bird. In the shop on our way out I proudly bought, and wore, a grey T-shirt with a beautiful image of a Peregrine on the front. Little did I know that seven years later I would begin studying this impressive bird.

In my infant school we had a Barn Owl flown around the hall – and without speaking a word we listened, but didn't hear a thing, as the owl's softened feathers allowed it to fly silently. During my time at first and middle

Figure a My childhood collection of Greenfinch feathers I found at my middle school.

school we had a few visits from people bringing along owls and falcons, and in my middle school we had a day learning about falcons and eagles. I remember having special access to spend time with the dozen or so birds at lunchtime, when they were relaxing on their perches on our school field. I just sat and watched, glued to their every movement as they preened and rested. And during the summer holidays my parents often took us to the Hawk Conservancy Trust in Andover – a chance to watch falcons swooping at speed past our heads, George the vulture thinking about the possibility of flying (but never quite getting airborne!), and a Tawny Owl flying between posts, still begging like a baby.

But my interest (and indeed obsession) with Peregrines really began with Greenfinches. Perhaps not the most obvious link, but bear with me. When I was eight years old and at middle school in Epsom, Surrey, playtime was always a diligent affair for me. In amongst playing tag and other games, I would be scouring the school field for feathers or watching the birdlife nearby – I particularly remember collecting up Starling and Pheasant feathers. Close to the school building a row of cherry plum trees was home to numerous Greenfinches. I didn't see the finches very often, but I did find their moulted feathers in the leaf litter below – their brightly coloured yellow wing feathers fascinated me. I loved trying to match them up in the order they would be found in the wing itself. I would pick up primary, secondary and tail feathers from male and female finches, keep them safe, and later set them out on cotton wool (Fig. a). I did the same with Mallard, Starling and Black-headed Gull feathers, and gradually built up a large feather collection during my teenage years. In 1992, I remember the excitement of visiting my secondary school for the first time and finding Lapwing feathers on the school field. These were a new feather species for me, and I realised Lapwings were using

Figure b The prey remains laid out in my student room in October 1998.

the field to rest and moult in the summer. I kept the feathers in my rucksack and took them home to add to my collection.

Throughout our lives, mentors can be important and indeed pivotal in how we decide to shape and evolve our careers, hobbies and relationships. I was very fortunate to have teachers (Mrs Sharman, Mrs Brewis and Miss Wood) and peers who encouraged and supported my love of birds. Two people in particular, John Tully and Nick Dixon, have supported my Peregrine studies during my time in Bristol. John sadly passed away in 2012, but during the 14 years I knew him he was a great role model and someone who gave me the opportunity to take on the wonderful work he had been doing on Peregrines. It was back during a very rainy October 1998 that John took me to Broadmead in Bristol and introduced me to the Peregrines that were using a tall office tower block. That day I remember we found the half-eaten corpse of a Lapwing, various parts of Redwings and Fieldfares, the legs and wings of Golden Plovers, and much more – all the prey of the Peregrines. I laid them out in my student hall bedroom to study and photograph (Fig. b). John regularly met with me and took me up the University of Bristol's Wills Memorial Building where we checked all the nooks and crannies for prey remains – the best we found was the ring from a Black-headed Gull which had originally flown here from Lithuania. Previously John had also found rings from Dunlin. Without John, I would never have got into Peregrines in quite the same way. I also helped out with the Bristol Ornithological Club's Peregrine Watch which John helped oversee, keeping an eye on the pair nesting in the Avon Gorge.

Not long after I began realising the potential of studying Peregrines I remember phoning Nick Dixon from a student house I was living in on the steep hill of Constitution Hill in Bristol. Nick was rather cautious of me at first – a young student suddenly taking an interest in the diet of Peregrines. Fortunately I met with Nick the following spring in Exeter and we had a stroke of luck. The Peregrines there had just eaten something very exciting – a Roseate Tern. This particular tern had a silver identification ring on its leg. It had been ringed as a chick in a nest on the island of Rockabill near

Dublin three years earlier. That December I visited Nick and his family on the edge of Dartmoor, and we went through various skulls and feathers from a gutter clearance of a church in Exeter – the remains included the skull of a Noctule Bat. Thirteen years later, I continue to help identify the prey of the Peregrines Nick studies in Exeter, and we have written various papers together. Nick also keeps tabs on all the urban and industrial Peregrines around the UK, and was involved with research on the impact of Peregrines on racing pigeons in the late 1990s. By working with Nick, we have been able to shed light on the nocturnal habits of Peregrines and liaise with others in the UK, Europe and the USA to amalgamate further evidence of night hunting. Nick Brown and Nick Moyes, who run a Peregrine project in Derby in England, have also provided support and evidence in the form of video footage of the Peregrines bringing prey back at night. The Derby Peregrines have become popular all around the world and the city even has a pub and a hotel named after the falcons.

In 1999 I received a small grant from the Millennium Awards, given by the Heritage Lottery Fund. With support from Chris Sperring (Hawk and Owl Trust) and Stephen Woollard, who supported this and other projects out of Bristol Zoo Gardens, I was able to really kick-start my work on Peregrines and take the project into schools and local communities. It was thanks to this award that I was able to take my research on Peregrines to a new level.

In 2007 I visited Poland to attend the 2nd International Peregrine Conference. This was a brilliant opportunity to meet other like-minded people working on Peregrines all across Europe and in other parts of the world too. From this conference a book, *Peregrine Falcon Populations* by Janusz Sielicki and Tadeusz Mizera, was produced, the most comprehensive amalgamation of papers and science about Peregrines in the twenty-first century. The conference also gave me a chance to network, publish more of my work, and discover more about Peregrines, particularly those living in urban areas.

The Peregrine world is now a very different place compared to when I first came to Bristol as a student in 1998. If I were coming to Bristol in 2014 I would be arriving in a city (and country) with many more Peregrines, and people studying them, than 16 years previously. Since I began studying Peregrines in 1998, there has been an incredible rise not just in the number of Peregrines using towns and cities across the UK, but also in the number of people watching and enjoying the falcons, and in those actively showing people Peregrines via nest boxes, web cameras and ringing chicks.

Why write a book about urban Peregrines?

In the late 1990s, a Peregrine in a town or city was a rare sight and there were just a few people watching or studying them, such as Nick Dixon in Exeter, Graham Roberts in Chichester and Brighton, John Tully (and later myself) in Bristol and Dave Morrison in London. Coming from the south of England, the only Peregrines I hoped to see were those wintering on the Kent and Sussex coastlines. When I came to study at the University of Bristol, I was both amazed and excited to learn that Peregrines were living in the city centre. Since then, Peregrines have become common all across southern England and while there have been declines across north and west Scotland, north Wales, parts of northern England and Northern Ireland, there are more Peregrines in cities and towns than ever before. As the species recovers across Europe and other parts of the world, it has continued to inhabit tall tower blocks and cathedral-type buildings in urban areas.

As Peregrines have become more widespread over the past 15 to 20 years, the wealth of information we have about them has also increased. But up until now, there has been nothing that specifically deals with urban Peregrines.

With new technologies and better accessibility, urban Peregrines have enabled us to glimpse more into their lives – it is a far cry from walking miles in open, rocky country to monitor a nest, although gleaning information from rural Peregrines is just as vital. Studying Peregrines in cities has enabled us to discover new details about their lives, including what they eat, when they hunt, and how they behave.

Throughout this book, I will be exploring the increasing use of urban areas by Peregrines, and what we are learning about their behaviour and lifestyles.

The book provides an in-depth view into what to expect from urban Peregrines, mainly in the UK but also in other parts of the world. It explores how to study them, interesting and unusual behaviours, and a perspective on the bigger picture, looking at how Peregrines will fare in a complex future with increasing urbanisation and proximity to people. It is also a chance to discover where to look for them and how to spot one of these magnificent raptors in a town or city.

Ed Drewitt

CHAPTER ONE

The Peregrine

I WAIT IN ANTICIPATION – I can see the parents sitting up on a pylon looking down at me in a yellow buttercup field full of inquisitive young bulls. We are by a cliff overlooking a railway and on the edge of suburbia. Meanwhile, climbers under my supervision and Schedule 1 Licence bring up a special package from the cliffs below. As I put my hand into the large cotton bag I smell a unique, musky scent which brings back memories of my contact with these birds last year. I was expecting three but instead we have four soft, large and very warm, fluffy Peregrine chicks. I like surprises, and sometimes you have one chick more than you expect. I quickly put a metal identification ring on one leg and a light blue colour ring on the other leg of each of the chicks. I then measure various parts of their head and legs, take their weights, and get them back to the climbers to settle them back in their nest, now with some colourful bling to show their parents. The colour rings will help us discover more about each bird during its life away from the nest. We leave promptly, and celebrate another successful and privileged opportunity to see Peregrines so closely and intimately.

The Peregrine, or *Falco peregrinus* as it is known in Latin, is a bird of stealth, a falcon of speed, and a hunter designed to kill (Fig. 1.1). For 6,000 years or more since falconry first became part of human culture the Peregrine has been a popular bird with people. However, it has also had its fair share of bad press, persecution and misunderstandings. Despite this, the Peregrine remains a popular bird with the public. Websites showing Peregrines live at nest sites from various locations around the world are increasing, while only the odd hummingbird, Osprey *Pandion haliaetus* or Northern Goshawk *Accipiter gentilis* may be lucky enough to be broadcast around the globe in such a fashion. And most spectacularly of all, the Peregrine has come back from the brink of extinction – not just in the UK, but also across Europe,

Figure 1.1 The Peregrine Falcon – now an urban convert (Sam Hobson).

North America and other parts of the world – after the devastating effects of chemicals used in farming and the countryside in the middle of the twentieth century. The Peregrine has returned with a vengeance in many areas, not only exploiting the countryside from which it disappeared, but also venturing into our towns and cities to become a truly urban species.

Recovering from around 385 pairs in the UK in the 1960s, the Peregrine had increased to 1,437 pairs and 1,530 occupied sites in the last survey conducted by the British Trust for Ornithology (BTO) in 2002. However, the overall population will be much greater that this 12 years on, despite declines in some parts of Britain. More recently Nick Dixon, who keeps track of urban Peregrine territories in the UK, has recorded at least 100 pairs living on made-made structures and at least 50 pairs in city centres (though not all may be breeding). Additionally, some Peregrines may go undetected depending on the surveys and type of data collection techniques used. Monitoring work by the Welsh Raptor Group has recorded more Peregrines compared to the BTO in the same areas in south and central Wales due to differences in criteria of data collected.

Monitoring of Peregrines in the UK has been ongoing for decades. Recent distribution maps (Fig. 1.2) from the Bird Atlas 2007–11, a joint project between BTO, BirdWatch Ireland and the Scottish Ornithologists' Club, show the whereabouts of Peregrines during the summer and winter between these years, and how their presence across the countries has changed over the past 40 years.

While Peregrines are doing relatively well in the UK, populations are very much still recovering in nearby countries such as Finland and Poland. In many European countries, captive-bred Peregrines have been released into the wild to increase and maintain the wild populations. In the Czech Republic in the mid-1990s there was only a single breeding pair in the whole country – now there are around 30 pairs thanks to increasing immigration from other parts of Europe. Even in Sweden, where much research has been conducted on Peregrines, there were still only 59 pairs in 2007. In Finland in 2012 there were up to 292 pairs, a vast improvement from just 30 in the 1970s, but still relatively low compared to the UK (over 1,400 pairs) and Germany (over 940 pairs).

Peregrines are found throughout the world, and aside from Antarctica and the Sahara Desert, they are able to live in most countries, islands and continents. Peregrines are also found in towns and cities all around the world, and many of their behaviours discussed in this book are reflected in individuals living in cities poles apart, from Australia to Argentina, Canada to Colombia and Iraq to India.

Figure 1.5 Peregrines are just as at home on a concrete office block as they are on a natural cliff or crag (Sam Hobson).

of the North American race make it to Scotland or Ireland as they move south through Greenland, Iceland and the UK instead of North America. Those individuals coming to the UK and western parts of Europe from further north in Europe will also be of the *F. p. peregrinus* race.

Why is the Peregrine found worldwide, while other similar species are not?

The Peregrine will live almost anywhere in the world, aside from those habitats which are just a little too extreme for it, such as the polar regions, deserts, and the islands of New Zealand and Iceland. The larger Gyr Falcon replaces the Peregrine in the Arctic regions of the world, while in the deserts you will find the Lanner *Falco biarmicus* and Saker Falcons, species better suited to the hot, dry conditions and lack of water – for example, the Saker's kidneys are more efficient and better able to conserve water than those of the Peregrine. Elsewhere, as long as there is food, mainly other birds, the Peregrine can survive in habitats ranging from mangroves to rocky coastlines, and cities to jungle forests.

The Peregrine is a very adaptable bird of prey, and will eat any type of bird that it is able to catch safely or bring to the ground and kill. Large gulls are the exception, and although they may be hit or attacked during the breeding season in territorial disputes they are rarely tackled in a full body grasp and killed. Falconry birds used at rubbish tips to control the presence of gulls are usually Peregrines which have been cross-bred with Sakers to produce a bigger bird with more speed and a more effective presence. These hybrids are able to bring the larger gulls to the ground and kill them.

Another key to the Peregrine's success relates to its ability to nest in a variety of habitats, including cliffs, trees, raised ground, buildings, and other man-made structures, from pylons to lighthouses. Additionally, it is able to travel long distances to follow its food when the seasons change. The Peregrine's ability to catch almost anything which flies, and its range of different migration routes between breeding and wintering habitats, means it is able to survive across more parts of the world than most other birds of prey. It shares similar global success with the fish-eating Osprey. Meanwhile, most other raptor species which breed in parts of Eurasia or North America may only winter in specific parts of Africa, South America or other parts of the world, and are more vulnerable to changes in habitats along their migration routes or in their wintering grounds. They may also be more specific in the range and type of prey they eat. Examples include many of the smaller falcons such as Eleonora's Falcon and the Hobby.

Figure 1.6 A Peregrine looks back through the stonework of a church (Dave Pearce).

What makes the Peregrine so successful in urban places?

The Peregrine has evolved adaptations that allow it to be resilient, robust and successful in a variety of habitats, including towns, cities and industrial locations. Tall buildings such as concrete office blocks, ornate limestone and sandstone cathedrals and churches, bridges, pylons, industrial buildings and telecommunication towers provide the ideal locations for a Peregrine to roost, breed and feed (Fig. 1.5). Just like natural rocky crags, sea cliffs or man-made quarries, these buildings provide a safe refuge at height, where the birds can see all around and make a quick exit to chase prey or see off an intruder. A gargoyle-clad church may look aesthetic and be an important place for worship, but to a Peregrine it is a rocky cliff overlooking parks, water bodies, and other buildings (Fig. 1.6). This is the Peregrine's world, where tens of metres high a quieter environment prevails, away from the noise and bustling activities of people below. And the resonating sound of nearby church bells is of little distraction – the falcons just get used to it.

The urban landscape provides a mosaic of habitats, from rivers to parks and woodland to rooftops. In turn, this variety of places provides a rich abundance of food for the Peregrines in the form of birds, and the occasional bat, rat or other animal. There is also easy access to the wider countryside, and an open flyway along which other birds are travelling, transecting across towns and cities on their journeys.

Figure 1.7 The Peregrine has adaptations that make it a natural-born predator (Sam Hobson).

Facial features and skull

Like other falcons, the Peregrine has a short neck, long, pointed wings, a hooked bill and sharp talons (Fig. 1.7). A closer look reveals some of the features that make it so successful.

Let's begin with the head of the Peregrine. The nostrils appear as holes at the base of the bill, surrounded by an area of yellow skin known as the cere (Fig. 1.8). Within each nostril lies a small knob-like structure. There doesn't appear to be much information on its exact function, but it is thought it may act as a baffle when the Peregrine is diving at fast speeds, reducing the flow of air entering the air passages. It may even act as a speedometer, helping the falcon to monitor its speed.

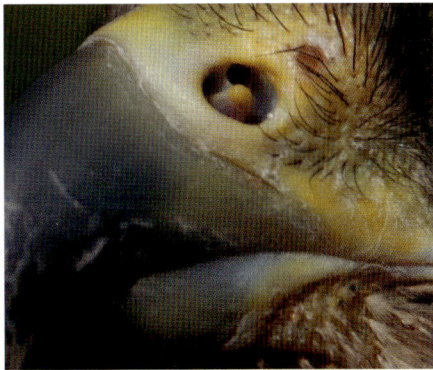
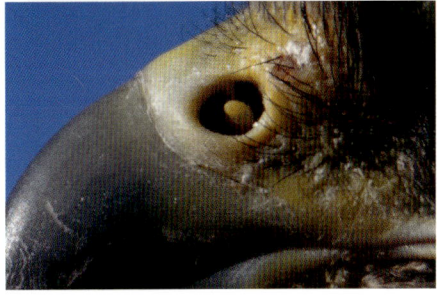

Figure 1.8 The cere of a Peregrine from two different angles.

The skin surrounding the Peregrine's eyes is featherless and yellow in adults and blue-grey in juveniles. Its eyes are dark in colour – a contrast to the Sparrowhawk or Goshawk, which are frequently confused with Peregrines but which have yellow or orange irises (Fig. 1.9).

If you look closely at a Peregrine you may notice it has a distinctive brow – this is an extension of the skull, a section of bone which is positioned over the eye and covered by skin. This brow is also seen in hawks, and is thought to act like a sunshade, reducing glare on a bright day. The bone itself isn't particularly strong but it may also help to support the large eyes. Additionally, the dark facial markings around the bill, eyes and front of the cheeks are thought to absorb bright light and reduce glare reaching the eyes.

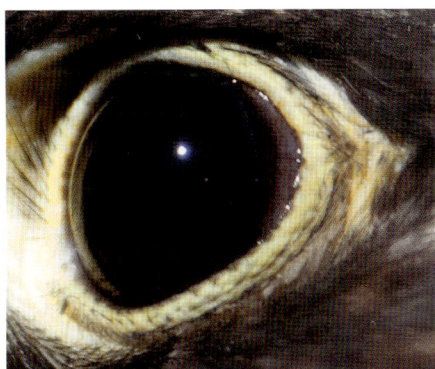

Figure 1.9 The dark eye of a Peregrine surrounded by yellow skin – here you can see the iris is in fact dark brown and not black.

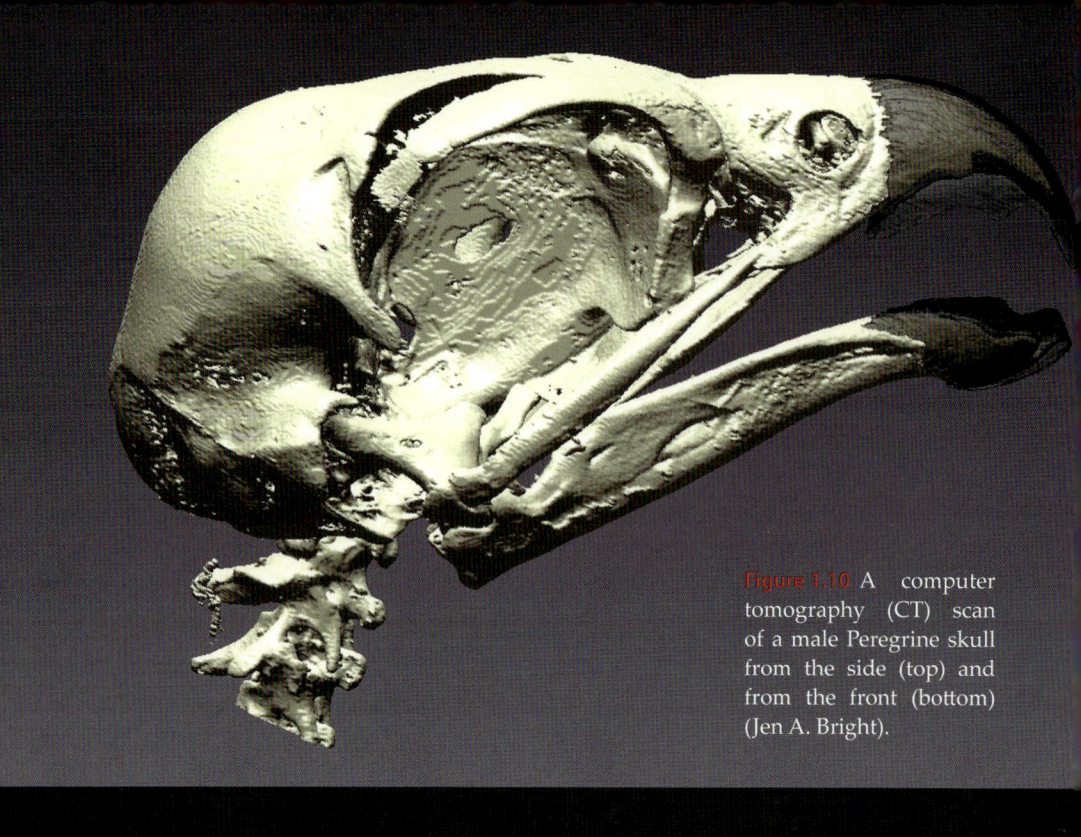

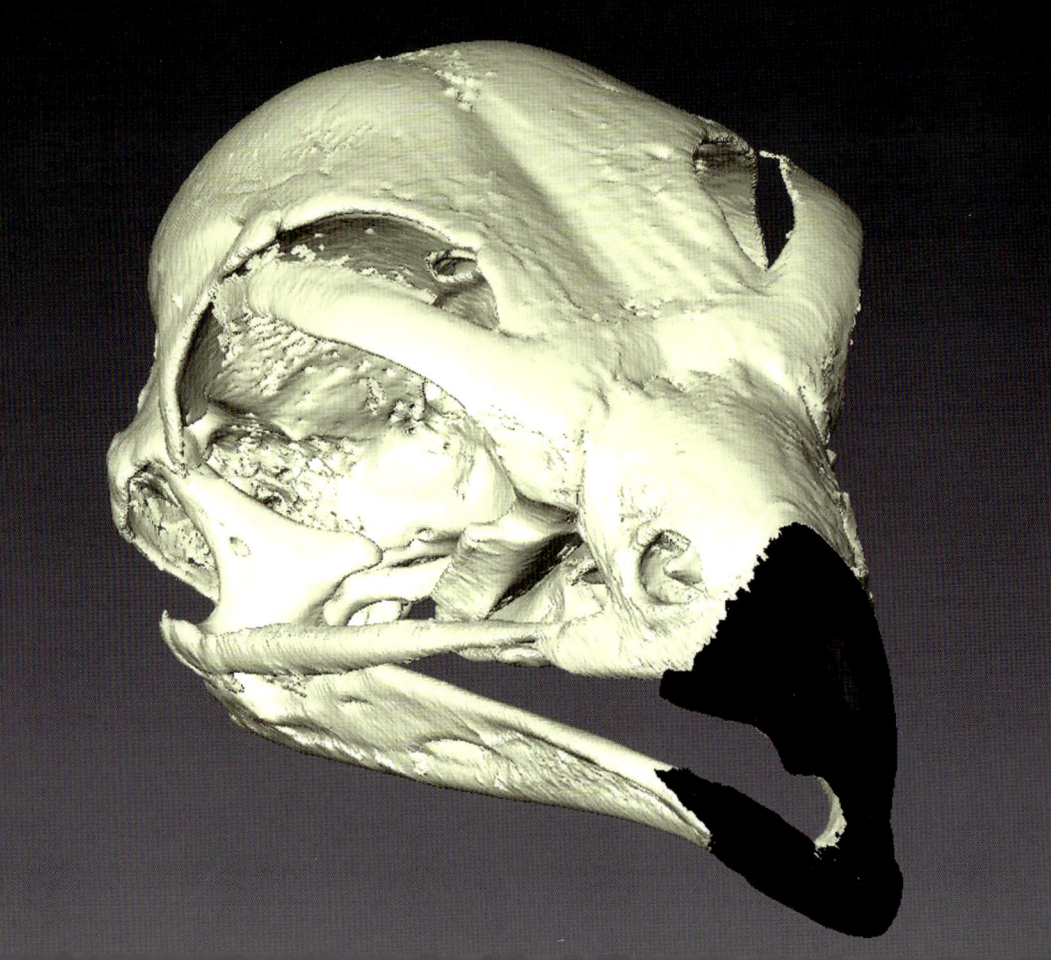

Figure 1.10 A computer tomography (CT) scan of a male Peregrine skull from the side (top) and from the front (bottom) (Jen A. Bright).

Figure 1.11 The big, forward-facing eye of a Peregrine (Ronald van Dijk).

At the University of Bristol, Dr Jen Bright has been studying the fine detail of raptor and parrot skulls with the help of Computer Tomography (CT) scanners to discover more about the muscles and bones which exist in both groups of birds. Little information exists on the anatomy of the skulls of raptors, and such scans allow the full architecture and structure of the skull to be examined without destroying it in the process. The research is revealing much more about how and why they look like they do. The scans of the 18-month-old male Peregrine skull in Figure 1.10 reveal incredible detail. What is immediately apparent on the cross section is how much empty space there is and how little of the skull is actually bone. The spaces are filled with soft tissue such as the brain and eyes, but reveal how light and hollow falcon skulls have become. The tomial tooth is clearly visible both on the bony part of the bill and also the keratin bill sheath which overlays it. In each eye socket, a sclerotic ring, made up of a circle of small sections of bone, would have been embedded within the eye itself as a form of support. Although absent here, thin, bony parts in the mouth and throat area of the skull are in fact bones which make up the hyoid apparatus and the tongue – as in all birds, the tongue contains bones which are covered in fleshy tissue and extend to the back of the skull.

What does a Peregrine see?

To be a successful top predator, Peregrines use a variety of senses. However, their sense of sight is the most pronounced, and essential for their survival. The eyes of the Peregrine are huge, forward facing, and make up 50 per cent of the volume of the cranium (Fig. 1.11). With so much space taken up there is little room for manoeuvre. The eyes are virtually fixed in place and so a Peregrine has to move its entire head to look around instead. With eyes like these, Peregrines are ideally suited for detecting prey at long distances, keeping focused when they are about to make their kill, and spotting prey during low light conditions, including at night.

The Peregrine is thought to detect movement and objects in similar detail to a young human. However, while the human eye has only one particularly light-sensitive area, known as the fovea, the Peregrine has two such areas. These sites are packed densely with light-detecting photoreceptor cells and are known as the deep fovea and the shallow fovea. They are positioned differently in the eye to provide the Peregrine with a variety of visual options. The deep fovea, which is thought to provide the most detail, helps Peregrines to see close objects, while the shallow fovea is used more for seeing distant objects.

When watching a Peregrine you may observe it looking straight at an object, but then tilting its head slightly to one side. These two viewing positions allow it to make use of both foveas and provide a more detailed picture of what it is looking at. If an object is a long distance away, Peregrines will look sideways at it rather than directly towards it, making use of the shallow fovea, which is most sensitive from the side. This can often be observed when a Peregrine spots another falcon, a Common Buzzard *Buteo buteo* or a kite flying overhead.

Figure 1.12 Unable to move its eyes like a human, a Peregrine tilts its head to get a better view (Dave Pearce).

Peregrines will also bob their heads up and down. It looks quite comical, but with little rotation of their eyes, these movements mean they get a better idea of what they are looking at and increase their field of view (Fig. 1.12).

Falcons are also excellent at fixing their eyes on to something while the rest of their body may be moving around. I remember watching a demonstration of a Kestrel at my middle school – its head was fixated on someone and the falconer was able to move the body of the falcon in a gentle circle while the head stayed in place. It makes sense – when you watch a Kestrel hovering, the head is fixed looking downwards, while the body and wings of the falcon may be moving around in the wind. It is thought that the way in which raptors use their eyes can have significant implications when it comes to changes in their environment, such as the appearance of wind turbines along important migration routes or in breeding territories. Vultures, for example, may find it very hard to see wind turbines in front of them as they have the most visual sensitivity below them, allowing them to look downwards to spot dead animals.

Like many birds, Peregrines are also able to twist their heads around 180 degrees or more, and while they may not practice it as often as owls, they do have some clever adaptations in the neck to allow them to do this safely (Fig. 1.13). A flexible neck is achieved by having double the number of neck bones compared to humans, while blood vessels are positioned more centrally in front of the spine, rather than on the side of the neck. To avoid cutting off the blood supply or suffocating, the arteries and other anatomy such as the oesophagus stay in place as the bird twists its head, meaning the head can rotate while allowing normal bodily functions such as blood flow and swallowing of food to continue without constriction.

Although the Peregrine needs to rotate its head to spot prey and other predators, rotating the head can cause an issue for Peregrines while in flight. In particular, greater air resistance (drag) is created when a bird decides to look to the side rather than straight ahead. If a Peregrine looks to the side by turning its head 40 degrees while in a straight stoop dive, the drag and therefore air resistance created doubles and will slow the bird down.

Biologist Vance Tucker has done some very neat theoretical and experimental work on the flight of Peregrines in wind tunnels to find out how Peregrines overcome this. He proposes that the falcons reduce this drag by diving along a logarithmic spiral path, keeping their heads straight, but with one eye looking sideways at their prey. Such a mathematical model shows that while a spiral path may be longer in distance, an 'ideal falcon' would

Figure 1.13 A juvenile Peregrine has a flexible neck and twists its head up and round without any undue effect (Dave Pearce).

reach its prey faster because its straight head would reduce drag. In reality, this does seem to be what happens – by tracking Peregrines in a stoop dive towards thrush-sized prey, Tucker has observed the falcons with their heads straight and following a curved path.

Seeing colours

What is really red? What is really green? We may all think we know what green is, but how we perceive colour as an individual may be different to how someone else interprets it. We are not able to assume that what may seem dark green to one person is the same hue to another person. The same can be applied to other animals. We used to assume that all animals, including birds, saw the world as we do. But we now know that animals such as birds and insects can also see colours which humans cannot. Humans are trichromatic, or sensitive to three colour pigments, red, green and blue, which are detected by three cone photoreceptors, the cells in our eyes which detect colour. Many birds, on the other hand, are sensitive to more colours as they have five cone photoreceptors (they are pentachromatic), and are able to see images and colours in far greater detail than humans. Birds are also sensitive to near-ultraviolet or ultraviolet light, which may be enhanced by the presence of oil droplets embedded within the five photoreceptors. These act as filters, only allowing certain wavelengths of light to be detected. In turn they make the avian eye more sensitive to light across a broader spectrum compared to humans.

Light waves are measured in nanometres (10^{-9} m) and as humans we are able to see light at wavelengths between 400 nm (blue-violet) and 700 nm (red). Between about 10 and 400 nm, light occurs in the ultraviolet (UV) range. One of the cone photoreceptors in a bird's eye has a maximum sensitivity of 370 nm, which supports the evidence that birds can see in the UV range. Behavioural studies on different bird species also support this.

Seeing ultraviolet colours has a range of benefits for birds, including spotting berries against non-reflective foliage if you are a Blackbird *Turdus merula*; detecting lakes and rivers from high altitude when on migration if you are a duck or wading bird; and seeing fish in the sea if you are an Arctic Tern *Sterna paradisaea*. It may also help birds to see certain feathers or plumages which reflect light in the ultraviolet range. While these colours may be invisible to humans, for many bird species they can provide important signals relating to health and display. The subject of ultraviolet light detection in birds has been a fascination of mine since I was a student. As a zoology undergraduate at the University of Bristol I decided to do my final year project on the ultraviolet light reflected off the iridescent blue-coloured patches (the specula) on the secondary feathers of male and female Mallard ducks *Anas platyrhynchos* (Fig. 1.14). I was blown away by the fact that Starlings, Blue Tits *Cyanistes caeruleus* and Zebra Finches *Taeniopygia guttata* use colours invisible to the human eye to decide whom they would like

Figure 1.14 The shiny blue iridescent feathers of a Mallard duck.

Figure 1.15 The Peregrine becomes the fastest bird in the world during a stoop dive. A falcon stoops in a spiral path rather than a direct straight line (Chris Jones).

As the Peregrine slices through the air, its death-defying speed is captured. The fastest speeds recorded and broadcast have been around 290 km per hour (180 miles per hour), which is remarkable, although up to 322 km per hour (200 miles per hour) for a very short amount of time is thought possible. One team in the USA, after film crews left, claim their falcon reached a speed of 386 km per hour (240 miles per hour). Sadly the filmed speeds don't usually get written up as peer-reviewed scientific papers, but nonetheless they make for fascinating viewing and provide some insight into just how fast Peregrines fly. Lloyd Buck, who flies falcons for television and film, has certainly recorded speeds of 290 km per hour (180 miles per hour) with his own falcon, although generally Lloyd finds the Peregrine won't do more than they need to. By flying his falcon Willow from the top of an 85 m (280 foot) crane and dropping food from the top, Lloyd found her stoop speed averaged between 97 and 113 km per hour (60 and 70 miles per hour), although on one occasion when she delayed her start, she reached 217 km per hour (135 miles per hour). The devices used to measure the birds' speed do cause some drag so slow them very slightly.

Peregrines certainly know how to fly, and when one is in a stoop dive it drops through the air with speed, grace and perfection. As the wings are drawn close to the body and almost closed, the bird forms a perfect

The Peregrine 21

Figure 1.16 The Buzzard (top), Sparrowhawk (middle) and Kestrel (bottom) can all be mistaken for a Peregrine (Pete Blanchard).

bullet-shaped aerofoil directed downwards towards prey, rivals, or a space where it simply practises its flying skills (Fig. 1.15). While the Peregrine may only reach these mind-boggling speeds for milliseconds, it allows the falcon to quickly move through the sky and catch up with an unsuspecting bird. Not all Peregrines favour the vertical stoop dive approach, and those living in deserts and the tundra are more likely to use fast, level, horizontal flights across these open landscapes.

As mentioned earlier, Vance Tucker's studies on the flight of Peregrines show that Peregrines don't simply drop in a straight, vertical line. Instead they drop at an angle so the Peregrine can both keep an eye on the target prey with precision and drop at a fast enough speed to close in and grab it.

During level flight the Peregrine is capable of flying at 45–60 km per hour (28–37 miles per hour); during an accelerated chase it may reach 100–130 km per hour (62–81 miles per hour); and in a stoop dive over 322 km per hour (200 miles per hour) is possible. Over a large area and distance, it is calculated that they could even reach 400 to 500 km per hour (249 to 311 miles per hour) in a stoop dive. Vance calculated that in a vertical dive, a 1 kg Peregrine will reach 95 per cent of its top speed in only 16 seconds, while travelling over a kilometre in the process; no wonder chasing prey amongst skyscrapers makes for a hazardous life! Such speeds cause Peregrines to experience g-forces far greater than what a person would experience on an adventure theme park ride or in a fighter jet. Despite its name, a g-force is not in fact a force but a measure of acceleration of an object, in this case a Peregrine, perceived as weight. It is expressed as a multiple of acceleration due to gravity (g). When a Peregrine closes its wings and drops like a lead balloon it is experiencing free fall and no g-force, but as soon as it pulls out of its dive and changes direction it experiences a tremendous g-force, between 18 g and 28 g. By comparison, during a roller-coaster ride we may experience between 3.5 g and 6.3 g, while a fighter jet pilot may experience 9–12 g. The Peregrine's size compared to a human means the blood has only a short distance to travel from the large heart to the brain, ensuring the falcon doesn't lose consciousness. With an average of 268 beats per minute at rest (in humans it is 60–100 beats per minute), the heart of a Peregrine is able to keep pumping the blood at high pressure to the major organs, especially the brain. This rate doubles during powered flight. Meanwhile, valves in the veins stop blood from moving backwards and draining from certain parts of the body and an efficient cardio-vascular system keeps the bird alive with no ill effects.

To help Peregrines deal with these high speeds and g-forces, they also have more durable and robust wing feathers than other falcons such as Kestrels.

Peregrines may also have similar adaptations to birds that fly at high altitudes. While wild Peregrines live on the cliffs of the Himalayas, Bar-headed Geese *Anser indicus* fly overhead at very high altitudes (6,500 to 7,300 m above sea level) through valleys in the mountains. Some even reach 9,000 m on occasions. They survive the lower oxygen levels at high altitude by having lots more blood capillaries in their muscles and a more efficient way in which the blood diffuses into cells. Peregrines may have similar but slightly more modest adaptations to reach high altitudes in the sky before a stoop.

Peregrine's size

When I spot a Peregrine in a town I generally think of it being crow-sized but more bulky. It is a relatively large bird, and females may in fact be as large as a small male Buzzard (Fig. 1.16), while female Sparrowhawks are only as large as a small Peregrine. Its relative, the Kestrel, is more obvious along motorways and open countryside. While it is often misidentified as a Peregrine, it is small in comparison, only 34 cm long and weighing between 190 and 220 g. In contrast, Peregrines are around 42 cm long and the larger females can weigh over a kilogram.

Sexual dimorphism

If you see a male and female Peregrine side by side there is a clear difference in their size (Fig. 1.17). The female, known as a falcon, is on average 15 per cent larger than the male, which is known as a tiercel. This is a common trait

Figure 1.17 A male Peregrine (below) and female (above) revealing the size difference between the two (Hamish R. Smith).

Figure 1.18 A changeover of incubation between the smaller male (left) and the larger female (right) (Dave Pearce).

amongst birds of prey, particularly *Accipiter* species such as the Sparrowhawk. The Peregrine sub-species *Falco peregrinus peregrinus* found across Britain and Europe shows particular sexual dimorphism. Females have a wingspan ranging between 930 and 1,000 mm while males range between 785 and 890 mm. Male Peregrines generally look slighter, smaller-headed and slim; females look broad-chested and fiercer, although this can be subjective and will vary between individuals (Fig. 1.18).

Such sexual dimorphism is often assumed to be a result of male and female Peregrines feeding on different size prey, with the female taking larger prey on average than the male. Additionally, this size difference is found more commonly in falcons that eat birds rather than those that eat insects. But perhaps surprisingly, both male and female Peregrines will target the same size prey and bring in similar prey species, although some may favour certain prey items over others.

So why are they so different in size? Rather than being directly diet-related there may be some form of non-random pairing happening, where larger females and smaller males are choosing each other as mates and successfully breeding together. One small study revealed large female Peregrines will pair up with small males, while smaller females will pair with more similar-sized males. In both scenarios, females are choosing relatively small-size males and hence any male offspring will be relatively small too. Either way, the larger females are able to produce larger eggs and chicks which may be more robust, more likely to survive post-fledging, and better at defending territories and defending against smaller male and female Peregrines. Although this has not been tested on Peregrines, Common Kestrels and Merlins have both been shown to select mates based on their tail length. Female Kestrels do this when there are a high number of males to choose from. Perhaps a longer-tailed male is bigger and therefore better and stronger at providing food and protecting her family.

Moult and preening

During the year of an urban Peregrine, the body plumage protecting it from the weather, allowing it to fly at tremendous speeds and keeping it camouflaged begins to fail. Feather tips become frayed or broken, and over time the feathers wear and become less efficient at keeping the bird insulated, waterproof and airborne. Therefore Peregrines, like all birds, need to moult their feathers.

Every year Peregrines will go through an annual moult to renew their wing, tail and body feathers. In its first year of life a Peregrine's moult is very obvious and a good way of ageing the bird (Fig. 1.19). The contrasting brown feathers give way to new grey ones, while streaky breast feathers are replaced with a paler plumage with horizontal bars (Fig. 1.20). Eventually 18 months to 2 years later the young bird becomes a sleek, slate-grey and white adult individual. For some juveniles this moult may begin as early as the March after they fledged, while for others it may start more towards the end of their second year. Most begin renewing their feathers at roughly a year old, during May and June. A British Peregrine which was colour-ringed as a chick in 2012 was already looking sleek in her adult plumage as early as in the August of the following year, so most probably began moulting early in 2013. She was just finishing moulting her outer primary feathers. However, more often you will see juveniles still showing brown feathers into the autumn.

Figure 1.19 As a first-year Peregrine moults there is a contrast between its brown juvenile feathers and its new grey adult feathers.

Meanwhile, adult Peregrines will tend to time the moult of their main wing feathers with their breeding cycle. The discarded flight feathers of Peregrines may be found below a church or other building during May and June, but sometimes as early as April. Unlike auks, rails, ducks, geese

Figure 1.20 The old juvenile breast feathers can be seen amongst the new white and grey breast feathers of a first-year Peregrine.

and swans, which moult their wing feathers all in one go and become flightless, Peregrines need to continue flying. They therefore replace their feathers gradually over a four- to six-month period, usually finishing in the autumn or early winter. There is a pattern to how the wing feathers are moulted. For example, Peregrines begin losing their middle primary feathers first. Gradually, each feather leading out towards the end of the wing is replaced in sequence, interspersed with the very inner primary feathers being shed. A similar pattern prevails for the secondary wing feathers. Female Peregrines may begin their wing moult after laying their third egg (if they lay this many eggs), but also have the ability to suspend moulting, usually when the young hatch. They want to be directing their resources and energy where they are needed most – catching prey and feeding chicks. Replacing feathers requires a lot of energy. It also means the birds are slightly less efficient at flying as they have gaps in their wings, changing the flow of air when in flight. By suspending moult, the new feathers can fully grow, leaving the females with fully functioning wings while they are caring for their chicks. The male Peregrines will usually begin their wing moult when the chicks hatch. The tail feathers of both sexes begin moulting anywhere between a week and 47 days after the first primary feather is shed. By the autumn most adult Peregrines will be in fresh, bright plumage ready to keep warm and protected during the winter months. Fresh plumage also means the falcons are in the best possible condition for hunting, and their smooth body profile is perfect for the most efficient and effective stoop dives.

To keep their feathers in good condition Peregrines will spend lots of time preening, carefully realigning their feathers, nibbling away at their shafts and gently pulling the feathers between their bills to re-zip any which have become tatty or split (Fig. 1.21). My favourite preening moments are when Peregrines begin to preen their legs and talons to keep their feathers and toes in order. They raise their legs up like drumsticks and carefully preen through

Figure 1.21 A juvenile Peregrine preens its wings and keeps its feathers in prime condition (Dave Pearce).

every small feather (Fig. 1.23). Peregrines will also find puddles or pools of water where they can wash and bathe. Peregrines have a preen gland at the base of their tail and will smear an oily substance across their feathers to help repel water, although when it has been raining really hard even this doesn't stop a Peregrine from looking wet and bedraggled. However, unless it is really torrential and persistent, rain usually just affects the outer parts of the feathers, and the inner, fluffier feathers tend to remain dry, keeping the Peregrine warm (Fig. 1.22). Peregrines living in the Queen Charlotte Islands

Figure 1.22 A juvenile Peregrine pauses to look around while preening its back and tail (Hamish R. Smith).

of British Columbia in Canada appear to have gone one step further. A fine powder, probably produced from the fine down body feathers fragmenting into tiny particles, is spread across their feathers, giving them further waterproofing and thus protecting them from the huge amounts of rainfall and mist this region receives each year.

Leg colour

Peregrines, including juveniles, have yellow legs (Fig. 1.24). There are lots of questions relating to the colour of raptor legs. Why do they have such brightly coloured legs and toes? The talons are usually clasping the prey to hold it in place – is it so they can differentiate between what they are eating and their own feet when everything is covered in blood and feathers? Is it an indicator of how healthy an individual Peregrine is? Are there other colours reflected from the leg scales of the Peregrine which only birds can see?

It is not fully understood why Peregrines have the yellow pigmentation – it may indeed be used as a way of showing other Peregrines how healthy

The Peregrine 29

Figure 1.23 An adult Peregrine nibbles and preens his talons (Hamish R. Smith).

Figure 1.24 The bright yellow talons of a Peregrine.

Figure 1.25 The yellow skin of an adult Peregrine may be an indicator of health (Dave Pearce).

and fit they are; that they are capable of catching plenty of food and defending a territory (Fig. 1.25). These honest signals of health are normally found in birds with feather colours derived from carotenoid-based diets. For example, in birds such as the Greenfinch *Carduelis chloris*, the healthier they are, the brighter the yellow in their wing and tail feathers. However, some studies do also show a link between bare skin colour and the health and fitness of a bird. For many birds it provides a way of expressing how well they are able to feed and how healthy they are.

While some questions remain unanswered, we do know that levels of testosterone control the presence of carotenoids in birds. Interestingly, male Kestrels have more brightly coloured legs around the time of mating, suggesting that colour is important in mate choice or the mating process, and may be regulated by testosterone. However, the presence of carotenoids does not appear to indicate how healthy a Kestrel is. Despite this, studies have shown that males with a good quality territory (in terms of size, habitat and prey) and the ability to bring in lots of prey have brighter coloured legs than those which do not. So perhaps a fit, healthy male Kestrel does develop deeper yellow or orange legs to attract a female partner. This may not translate directly across to Peregrines of course, but it is an interesting hypothesis to explore. Studies on the yellow cere colour of male Montagu's Harriers *Circus pygargus* have shown that the cere reflects lots of light in the ultraviolet range, part of the light spectrum invisible to humans but visible to birds and insects. The more orange the cere, the brighter the ultraviolet colours reflected, and the more healthy the harrier in terms of weight and condition. Further studies into mate choice are needed in relation to ultraviolet reflectance, but male harriers with brighter coloured ceres have been found to be breeding with early laying females and may therefore be snapped up quickly by partners looking for fit, healthy males.

Figure 1.26 Two juvenile Peregrines relaxing in Cheltenham showing their brown plumage and blue-grey bare skin (Dave Pearce).

Interestingly, while young Peregrines have yellow legs like the adults, their eye skin and cere remain a blue-grey colour (Fig. 1.26). The duller plumages and colours of younger falcons help to keep them cryptic and hidden. Their priority is survival and learning life skills rather than impressing a mate or providing for others – only when they learn to hunt for themselves and change into adult plumage do they develop a yellow cere and eye skin.

Age

Age is always difficult to ascertain without knowing when a bird actually hatched. However, with more studies tagging Peregrines from a known age, for example as a chick, Peregrine researchers are getting a better idea of how long they live for. On average Peregrines survive until they are 5 or 6 years old, though wild colour-ringed birds in Sweden and Germany, usually females, have lived up to the age of 17. Males have been found alive up to 16 years old in the USA and 14 years old in Europe. However, around 50–60 per cent will die in their first year, succumbing to the hazards of life as an inexperienced bird, and often starving to death. Once they have got through this year, their survival rate increases to 80 per cent or more. In Scotland, the Lothian and Borders Raptor Study Group has found that male Peregrines live an average of 8.8 years, and females 10.4 years.

CHAPTER TWO
What is an Urban Peregrine?

WHEN I LOOK OUT across the city of Bristol in England from a high building, I am always amazed how green it is. Not all cities may be as green as Bristol, but the fact is our towns and cities are diverse and not just concrete and tarmac. Bristol may not have as much diversity of life as ten miles out into North Somerset, but it still has a huge variety of organisms from lichens to bryophytes, beetles to moths, and birds to mammals. Although buildings don't suit most wildlife, there are some species such as Swifts and bats which in today's world need them to breed.

In 2013 a collaboration between 25 UK conservation and research organisations produced a report called *The State of Nature*. It was a comprehensive

Figure 2.1 A Peregrine against an urban backdrop (Sam Hobson).

Figure 2.2 Birds such as Common Terns (top) are at home in some towns and cities, such as London and Helsinki, while Wild Boar (bottom) roam urban areas such as Berlin.

review and assessment of just how wildlife, plants and other organisms, including those living in urban areas, are faring across the UK. With almost a tenth of the UK covered in towns and cities, and 80 per cent of people living in them, there is huge potential for everyone to get outdoors and experience nature on their doorstep. Just a bus ride or short walk away we are able to see unexpected species which have become urbanised, as well as those animals and plants that are less obvious or common. The report makes for both fascinating and eye-opening reading. While many of the facts reveal what we are losing, they also reveal what we have gained and explore how and why things have changed. This in turn provides solutions as to how various species can be helped longer term. While almost two-thirds of species found in urban areas have declined, there are some animals that are doing really well – and one of these species is the Peregrine. Over the past 10 years, many populations of Peregrines across Europe have had annual growth rates of 10–15 per cent. Peregrines have definitely been on the up (Fig. 2.1).

Cities often contain a mosaic of habitats such as a river, a variety of parks, open grassland, allotments, and much more. Many birds and mammals which we associate with quiet, remote places, from otters to Kittiwakes *Rissa tridactyla*, Common Terns *Sterna hirundo* (Fig. 2.2) to Goosanders *Mergus merganser*, and Red Squirrels *Sciurus vulgaris* to Wild Boar *Sus scrofa* (Fig. 2.2), can be equally at home in towns and cities around the world. And Peregrines are no exception. Where there is food and somewhere to nest, city life is the alternative life for the Peregrine.

The Peregrine is adaptable, opportunistic and fearless; the buildings are just an extension of its natural habitat of cliffs, moorland and estuaries. Buildings are often made of the same type of rock, such as limestone and sandstone, and even if they are concrete, the peace and quiet up high is a

Figure 2.3 An urban Peregrine flies past the windows of a high-rise office block (Sam Hobson).

world apart from the hustle, bustle and noise on the ground. The Peregrine is not alone in the raptor world – Goshawks, Kestrels, Hobbies, Buzzards, Sparrowhawks and various owls also make their homes in cities in Europe. But the Peregrine is the most obvious, bold and reliable to see.

Towns and cities offer an abundance of food for Peregrines, from pigeons to Blackbirds. The mosaic of habitats provided by cities, and their close proximity to the countryside, means Peregrines have a varied and changing menu throughout the year. While Red Foxes *Vulpes vulpes*, pigeons, rats and squirrels may often dominate people's thoughts or be the most obvious city dwellers, a variety of insects, woodland and garden birds, and small mammals such as voles are often still in abundance where there are green spaces. Urban Peregrines are looking for plenty of prey, open spaces for catching birds, and ledges where they can pluck, eat and store food (Fig. 2.3). Towns and cities may often show a decline in the diversity of species compared to the countryside, but many still support a huge range of species both in the suburbs and in the centre. Towns and cities are also slightly warmer than the countryside, and many birds survive better here during cold winters. This means that even on the colder days there is still plenty of bird food for urban Peregrines.

For Peregrines which live during the summer in the northern parts of

Figure 2.4 A Herring Gull is just as much at home in a city as it is on a remote coastal island.

Scandinavia, winter becomes a life of contrasts. Most, if not all, depart and move further south, leapfrogging over those that have bred further south and stay further north than the migrants for the winter. Many even reach northern and western parts of Africa, such as Morocco and Senegal. Whilst many will spend time on estuaries and lowland wetlands, others will roost in cities across Europe and become 'urbanites' for the winter.

The density of Peregrines varies around the world. Some of this is historical, a relic of their decline due to persecution and pesticides over the past 150 years, while for other locations it may be due to the type of habitat and the availability of food. For example, in Mediterranean countries during the hot summer months many birds may be sparse or hidden away in bushes. The Peregrine's relatives, such as the Lesser Kestrel, Common Kestrel, Hobby, Red-footed Falcon and Eleonora's Falcon, prosper instead with their ability to feed on small mammals, insects and reptiles, as well as on small birds such as hirundines and Swifts.

In North America, many Peregrines depart the continent and spend their winter across countries in South America. One Peregrine traced by satellite telemetry in the 1990s travelled from her breeding grounds in Alaska all the way to the wilds of Argentina for the winter. Her travels were made into a lovely illustrated book for families, encountering all the different places and habitats on her journey, including the urban city of Seattle, Washington.

Peregrines are reasonably new converts to urban living in the UK and across Europe, but their presence in cities in North America was established during the 1930s and 1940s when skyscrapers and other buildings became their new homes. A famous pair laid eggs on the Sun Life Building in Montreal, Canada, in 1938 and 1939 (including five eggs in one clutch); a pair first nested in New York City in 1943 while another was discovered in Philadelphia in 1946. The New York Peregrines have been well documented, commonly wintering there between 1949 and 1959. Sadly their population disappeared from 1961 onwards as the species declined due to the effects of pesticides. Their more recent comeback and success has been put down to the abundance of suitable ledges on skyscrapers and bridges in Midtown and

Figure 2.5 A Lesser Black-backed Gull is always on the lookout for a free, easy meal.

Downtown Manhattan, as well as the absence of large trees which impede their flight. Today 69 per cent of Peregrines in New York State breed on man-made structures.

Peregrines didn't totally avoid urban sites in Europe, however, and pairs were found nesting in old towers and castles in Germany in the 1950s. In London and the Home Counties, Peregrines were seen from time to time during the early 1900s, usually between October and May. During the 1920s, 1930s and 1940s they were sometimes spotted over the dome of St Paul's Cathedral, perching on Big Ben, or flying over, sometimes staying for days. St Paul's appears to have been a favourite place for urban-dwelling Peregrines even during the Victorian period.

As Peregrines have increased in towns and cities, so too have gulls and pigeons, both successful converts to urban living. I've realised over time that gulls and Peregrines have a lot in common – they are both very bold, adaptable birds, and are good at making the most of the resources they have around them. They are also similar in how they disperse (males stay closer to where they hatched), how long they live for (both are long-lived), and the variety of food they eat. And despite many people's hopes, Peregrines generally don't eat the large gulls such as Herring *Larus argentatus* (Fig. 2.4) and Lesser Black-backed Gull *Larus fuscus* (Fig. 2.5).

Since 2000, I have worked closely with Peter Rock on urban gulls – he's been studying them since 1980. When I visit roofs around Bristol each summer to help ring the gull chicks, you get a brief glimpse of what it must be like to be up high and nesting there. The roofs are like an archipelago of islands, often covered in small plants such as stoneworts, which are experts at growing on exposed places with little substrate. Gulls have shifted from being 'wild' birds feeding on fish and other marine life to becoming 'urban' converts feeding on human leftovers and litter on our streets, as well as natural foods such as worms and small mammals on local playing fields and rough pasture. We have discovered everything from earthworms to deer legs, moles to small birds, and kebab sticks to pizza both around the nests and sometimes regurgitated right in front of us (pizza, sandwiches and whole chicken breasts anyway!). The gulls terrorise the pigeons and unlike

Figure 2.6 A young Peregrine ducks down as a Herring Gull comes in to attack it (Hamish R. Smith).

Peregrines lack the talons and beak to quickly dispatch them. Instead, they turn the pigeons inside out like a glove, to gain the flesh inside. This is very similar to what they do to petrels and shearwaters on marine islands. The large gulls such as the Herring and Lesser Black-backed Gull generally leave the Peregrines alone, although they will still harass young birds, particularly around fledging time (Fig. 2.6). As both gulls and Peregrines have increased in cities, younger birds have been recruited into the breeding population at an earlier age, and recently we have seen first-year Peregrines trying to nest or help out their parents. It is clear that in the same time that gulls have ventured into our towns and cities to take advantage of our throw-away society and free roof spaces, Peregrines have also followed suit to take advantage of the glut of food, safe roost sites and various nesting possibilities.

Paolo Taranto has documented in Italy just how Peregrines naturally colonise our towns and cities. Peregrines don't usually just turn up in cities and begin breeding. There is a 'pattern of urbanisation', a behaviour which was very obvious in the UK during the early 2000s when Peregrines were beginning to colonise urban areas. During the early stages of Peregrines visiting towns and cities, just one bird may appear from time to time on particular buildings. In fact this one bird could be a number of different individuals all present at different times. However, in Paolo's study this was usually a male bird. Over time, perhaps the following two or three years, a second bird may appear. Both stay loyal to a site or to a few sites over the course of a year or two, but as a non-breeding pair. They may then attempt to nest – if a suitable nest ledge isn't available, they may nest in spaces where the eggs are vulnerable to rolling away into the gutter or away from the nest itself. If breeding is successful on a ledge or artificial nest site, then the pair may nest again or over time be replaced by new birds (Fig. 2.7).

This pattern of urbanisation is replicated all around the world. In the UK, Peregrines have been established now for over ten years or more in some places, but in areas where the species is still uncommon or only just arriving, this pattern is still taking place. Where breeding doesn't occur, it may be that

What is an Urban Peregrine? 39

Figure 2.7 A male Peregrine looks out from his favourite perch in Cheltenham. Once this individual had found a mate, the pair laid eggs on the bare roof where the eggs rolled into the gutter. A nest box has since been installed and successfully used (Dave Pearce).

the single bird or pair is using the location(s) as a hunting habitat and breeding elsewhere in a quarry or natural cliff 1 to 6 km (0.6 to 3.7 miles) away, or that the pair simply don't have a suitable nest ledge to even attempt to lay eggs on. In Worcester, England, a female colour-ringed Peregrine called Bobbin who was hatched in the city in 2009 has been using the cathedral during the winter months. However, she disappears during the summer period and has been relocated 35 km (22 miles) away in the neighbouring county of Shropshire, where she breeds.

Even where Peregrine populations are relatively low, in countries such as Hungary, the Czech Republic, Romania and Bulgaria, the species still turns up in cities. In Bulgaria, where surveys for Saker Falcons in the mountains instead revealed Peregrines, cities such as Sofia and Burgas see a few falcons arriving in winter after migrating down from the mountains. In Hungary there may be as few as 12 pairs breeding in the mountains, with many former sites now taken over by Saker Falcons. However, during the winter Peregrines from more northern countries pass through Hungary or settle in the lowlands for the winter. And the odd individual may therefore be seen in the centre of Budapest.

In the state of North Rhine-Westphalia in Germany, there is a huge industrial area from Duisburg to Dortmund which includes the cities of Cologne, Bonn and Düsseldorf. In this region 93 per cent of Peregrines nest on man-made structures. This is a very high percentage, and across the whole of Germany 37 per cent of the total breeding Peregrine population nest on buildings. This is a contrast to the UK where the figure is less than 10 per cent. During the breeding season, Peregrine researchers in this part of Germany are busy visiting nest sites and ringing chicks – there may be over 180 sites to visit in the space of a month, leaving the scientists with only 26 minutes per site to ring! In 1970 there were no Peregrines in this region, but

30 years ago captive-bred Peregrines were released into the wild. Now it has one of the densest populations of Peregrines in Europe and perhaps the world, with 3 pairs per 1,000 sq km (621 sq miles).

Why are so many Peregrines nesting on man-made structures in Germany? In the absence of other suitable nest sites, many of the buildings have artificial nests installed so the Peregrines have somewhere to breed. Nest sites have been monitored to ensure persecution and loss to predators such as Eagle Owls *Bubo bubo* (Fig. 2.8) or other natural causes is minimised. Studies there also show that Peregrines often switch to nesting on buildings over other types of nest sites, and the traditional cliffs and quarries in this region are only used by 7 per cent of Peregrine pairs.

Figure 2.8 Eagle Owls may seem an unlikely predator of Peregrines. But at the nest, chicks and incubating adults can be at risk.

With Peregrines increasing across many parts of their range, and so many people taking an interest in and researching their habits and behaviours, there is a great opportunity to better predict how the species will populate and survive in urban areas over future decades. While the density of Peregrines in the UK may not be quite at the level of North Rhine-Westphalia in Germany, the UK certainly has one of Europe's largest populations of Peregrines overall. Already Peregrines are beginning to show signs of nesting in closer proximity in the UK, taking up a wider range of nest sites, and using ledges and sites which up until recently may have been less desirable. With some of the best sites taken, new recruits in the breeding population can't afford to be fussy!

Chapter Three

How to Spot a Peregrine

When I used to sit at my office desk in Berkeley Square in Bristol, I frequently looked up and saw a Peregrine dashing past. They don't fly particularly fast in level flight, and their size and fluttery wing beats instantly trigger my 'jizz' button, the ability to identify a bird just by seeing something about its shape or behaviour. That is all very well when I have been studying Peregrines for a long time, but what if you have never seen a Peregrine before?

In an urban environment, spotting Peregrines can take time, even if you know where they should be on a roof or gargoyle – but once you get your eye in they can be quick to find. In flight Peregrines look large – about crow-sized,

Figure 3.1 A male and female Peregrine at rest on a church in Bath. From a distance the birds almost disappear and look like part of the stonework (Hamish R. Smith).

Figure 3.2 A Peregrine perched on an old brewery in Bristol (Sam Hobson).

and bigger than a Sparrowhawk or Kestrel, with some females as large as a Buzzard. The smaller male Peregrines may be a similar size to a female Sparrowhawk. Peregrines fly with shallow wing beats, which give a 'fluttery' appearance. They have a broad, barrel-shaped torso, and appear shorter-tailed than a Kestrel, which has thinner, shorter wings.

On a building, Peregrines usually stand upright and look hunched, small-headed and short-necked. Peregrines have evolved to be camouflaged against cliffs, and their colours also help to keep them hidden against buildings, whether they are made of concrete or natural stone (Fig. 3.1). On closer inspection, the dark hood and moustachial stripes of the Peregrine can be seen along with the contrasting barred markings (horizontal lines) on the breast (Fig. 3.2). Juvenile Peregrines are brown-backed and have a cream-coloured breast with streaks running down it as vertical lines. They have a brown head, a pale collar which doesn't extend round to the back of the neck, and dark moustachial stripes. Depending on the location, adult and juvenile Peregrines may look lighter or darker, and have varying degrees of white or cream on their breast feathers.

Often Peregrines are so well hidden that only by looking closely at the stonework may you notice an unusual shape, particularly if they are perched on a gargoyle. Any irregularity on a ledge, spire or mini-spire of a church, cathedral or other building will often reveal a Peregrine at a known site (Fig. 3.3). They have their favourite perching spots too. My study birds in Bath have predictable perches on the limestone carvings which they prefer to stand on, looking out across the Georgian architecture of the city. Sometimes they may be hunched up and sitting very close to the concrete or brickwork of a building, making them even harder to see.

Peregrines are noisy too, making a variety of shrill 'kekking' sounds alongside drawn-out 'mews' and cries. Describing sounds in words that others can decipher is never an easy task, but searching for Peregrines and their calls in an online search engine is a good way of hearing some of their sounds. The young birds are very vocal, particularly around fledging, and keep up their long, drawn-out cries for a long time. One of my favourite Peregrine calls is their 'ee-chupping' sound, a succession of chicken-like clucks interspersed with the odd grunt. This is used by a pair when they are courting and being intimate, and also by the female when feeding her chicks. Until recently it has been a sound that was rarely heard, but thanks to web cameras and microphones it is now more familiar to those watching courtship and nesting behaviour of Peregrines online. More likely to be heard are long, drawn-out wailing calls used to indicate wanting something, and creaking calls used during courtship and display.

Figure 3.3 Two Peregrines sitting on an office block (Matt Allen).

If visiting a town or city for the first time, it is best to look for Peregrines on tall office blocks, churches and cathedrals. Often looking at the base of the buildings will reveal if any Peregrines are around – look for pellets, feathers and white droppings, known as mutes. The latter are thick and white, not diluted or opaque, and form oval shapes up to 4 or 5 cm (1.5 or 2 inches) in diameter on the ground, although this will vary depending on the height at which the bird was standing when it defecated. Often their mutes form

Figure 3.4 A Peregrine retrieving a pigeon from its favourite cache in a gutter. Prey such as pigeons are so big they often overhang gutters and are a sure sign a Peregrine is in residence (Gary Thoburn).

streaks down the side of buildings where they have been perching. The next step is to look up towards the gargoyles, posts and tops of pillars to see if a Peregrine is perched. Equally, it may be possible to see the leg or wing of cached prey hanging over a ledge, or out of a drainage tank or gutter (Fig. 3.4). I remember visiting the town museum in Dorchester, Dorset one day and having a feeling this would be a good place for Peregrines. Before I knew it, I had found some feathers from Eurasian Woodcock *Scolopax rusticola*, Fieldfare *Turdus pilaris* and Feral Pigeon *Columba livia* below the next-door church. And it wasn't long before a Peregrine flew overhead. I found out later that up until this point, Peregrines had been largely overlooked in the town.

As illustrated by this Peregrine, if the species is around, individuals will eventually reveal themselves at some point, often flying round or onto a building, and calling. I didn't have to wait long in Dorchester, although if Peregrines are out hunting you may have to wait some time. Back in Bristol, my old office was very close to the University of Bristol's Wills Memorial Building – I was amazed how regularly I would hear and see Peregrines drift past, usually in the direction of the building. While they don't nest on the building, it is the tallest landmark around. Juvenile Peregrines often use it post-fledging, while adult Peregrines use it throughout the year, particularly during the winter. I always reckon I am more likely to see a Peregrine in Bristol than I am a Sparrowhawk, Kestrel or Buzzard!

When is a good time of the year to spot Peregrines?

Peregrines are visible all throughout the year in towns and cities, although locations which are in the early stages of the pattern of urbanisation may find there is an absence of Peregrines between April and July, when individuals move off to breed elsewhere or travel around the region or country.

Late December sees local breeding urban Peregrines become more conspicuous as they begin courting, displaying and copulating. Their flying antics are more obvious over breeding sites, and their vocalisations may become louder and more regular. Peregrines are also particularly visible when they have large or fledging chicks to feed. Once a whole family fledges, up to seven Peregrines could be in the air at any one time if the parents have a brood of five chicks.

Figure 3.5 A juvenile Peregrine has just fledged but become grounded. In this case the bird was healthy and put back to a higher perch (Dave Pearce).

If there are many non-breeding Peregrines in a region, then there may be regular interactions between these and breeding pairs as they pass overhead. Graham Roberts who studies Peregrines in Chichester, West Sussex, has noticed that when spending time watching the local pair it is common to see an intruding bird passing through or investigating the site at least once a day. The resident birds quickly see it off, although sometimes an intruder may decide to stay and try to take over his or her rival's position, leading to a fight and perhaps injuries.

What to do if you find a Peregrine

Peregrines often find themselves grounded. Often it is a young bird which has just fledged and may not quite be able to fly properly or is too heavy to move anywhere! In some locations, you may even wake up to find a Peregrine perched on your balcony – I remember once watching a recently fledged Peregrine sitting on the lion enclosure at Bristol Zoo. And sometimes residents in London may open the curtains one morning to find a juvenile bird sitting on their balcony railings.

During the fledging season, juvenile Peregrines may stand on the ground or perch on roofs very close to people (Fig. 3.5). They are very tame in the first few days after fledging and if left somewhere elevated they will be fine left alone. If they don't attempt to fly away from the ground and are vulnerable to cars, dogs and people, then some quick intervention by the site's local watch group (if it has one), the Hawk and Owl Trust or a Peregrine ringer may be needed. Usually just putting them on a nearby roof close to the nest site or even back in the nest will work – the latter would need to be done by someone who has a Schedule 1 Licence for the site. Sometimes, though, a young bird may genuinely be injured – a fall from the nest can occasionally cause spinal or other injuries, or its injuries may have been inflicted by a person or other animal. In this case the bird may need further observations or treatment by a wildlife hospital or falconry centre who specialise in bird care, or by the Royal Society for the Prevention of Cruelty to Animals (RSPCA). If you are unsure, then it is good to get advice from one of the above or someone local who watches and oversees the birds and nest. It is worth being aware that if the bird is not able to make a full recovery it may need to be euthanised, even if it seemed fit and well, as further explorations can reveal hidden internal injuries.

If a person has caused the injury on purpose and therefore a crime has been committed the police would need to be informed. Some police forces have more resources than others to deal with wildlife crime – either way, it is worth persisting to get a crime number and ensure the instance is taken seriously. It is also worth informing wildlife conservation officers in the local authority as well as contacts in government advisory bodies that deal with the natural environment (in the UK, these are Natural England, Scottish Natural Heritage, Natural Resources Wales and Northern Ireland Environment Agency). The more official people aware of the crime, the more likely it is that it will be taken note of and action taken. In the UK, you can also inform the Investigations unit of the Royal Society for the Protection of Birds (RSPB) and the Partnership for Action Against Wildlife Crime (PAW).

Chapter Four

A Year in the Life of an Urban Peregrine

When discussing a year in the life of an animal it is often easy to begin in January and work your way through the calendar year. However, January isn't necessarily the most interesting or appropriate moment to explore a year in the life of an urban Peregrine, especially when you consider that Peregrines are found across the world and will be nesting at all times of the year in different places.

July	Aug	Sept	Oct	Nov	Dec	Jan	Feb	Mar	Apr	May	June

Late winter through to early March: **courtship and display**

Mid-March to early April: **laying eggs**

March to end of April/early May: **egg incubation**

May and early June: **chicks in the nest**

Mid-June onwards: **chicks fledge and disperse**

July to April: **non-breeding Peregrines or breeding birds from rural sites sometimes appear in towns and cities**

Year round: **breeding Peregrines stay in towns and cities**

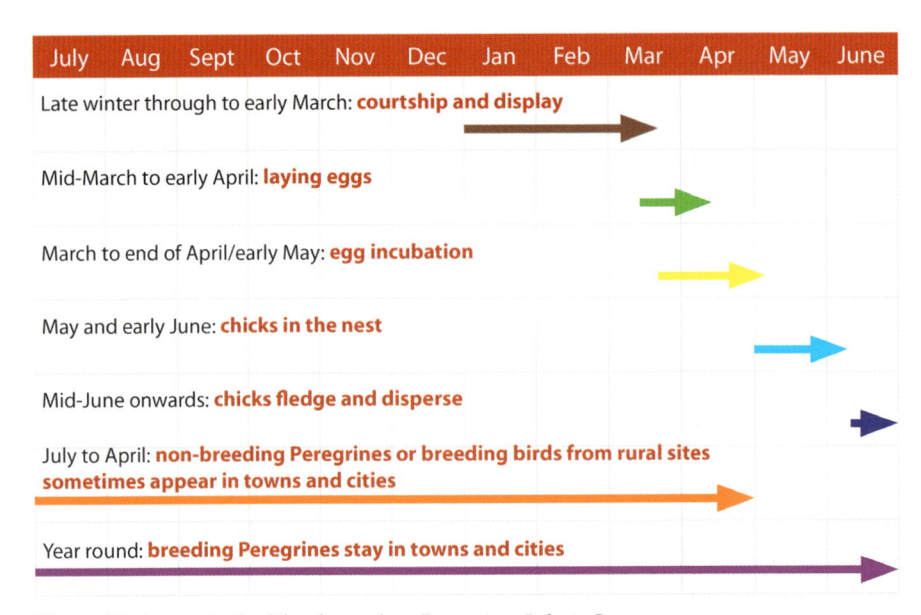

Figure 4.1 A year in the life of an urban Peregrine, July to June.

Late summer

I will therefore begin in late summer, around July and August in the UK (Fig. 4.1). This is a time when towns and cities which lack breeding Peregrines may find one or perhaps two individuals appearing and staying through to the following spring, usually around April. This is the process of the pattern of urbanisation as discussed in Chapter 2, which describes the behaviour of Peregrines appearing in urban areas around the world. Over a few years it may lead to breeding success if a suitable nest site is available. When I first came to Bristol, most Peregrines across the country were at the early stages of this model. Some pairs were nesting in cities such as Exeter and Chichester, but in others such as Bath, Peregrines were present mainly between July and April. It was a mystery where they went during the summer months – perhaps some went back to northern Scandinavia in early June, where breeding would commence once the snow and ice had melted. Or perhaps they moved off to search for suitable nest sites around the region. More recently, observations of individual Peregrines reveal that some move away and nest out of town in a cliff or quarry location, only using the urban location from July onwards. In other countries such as Hungary and the Netherlands, it is more common to have birds just wintering in a town or city before heading north to breeding grounds, or to a higher altitude up in the mountains. In the UK, Peregrines in rural locations generally stay close to their breeding sites, even if they have to move to a slightly lower altitude – the milder winters compared to mainland Europe mean food is still available in these locations and the birds don't need to move into towns and cities.

Autumn and winter

How many Peregrines visit urban sites?

Towns and cities are host to many more Peregrines during the winter months than you might imagine. Obvious buildings such as tower blocks or churches see many birds passing through – these nomads are called satellite birds. This is a time when younger Peregrines in particular become travellers, exploring the countryside and wandering away from where they were hatched. In Chichester, Graham Roberts, who has been studying the Peregrines in the city since the late 1990s, observes at least one satellite Peregrine passing over or interacting with the breeding pair during any one day. Sometimes they are individuals from nearby towns or cities, but often they are non-breeding birds checking out the potential availability of all the local nest sites.

Figure 4.2 The colour ring of a Peregrine is unique and allows researchers to find out who's who at known nest sites (Hamish R. Smith).

Usually we can tell different Peregrines apart if they have had a colour ring fitted on their leg as a chick (Fig. 4.2). Unless they are marked in this way it can be tricky telling individuals apart. However, when urban Peregrines are observed closely, particularly on web cameras, subtle differences in plumage and age can mean different birds are identified, revealing more than just the usual suspects. For example, in 2013 Worcester Cathedral was home to 10 or 11 different individual Peregrines – many were just present for a few hours or a day. One was a known resident adult female who was hatched at the site a few years earlier and often spent time there; one was a colour-ringed bird from Cornwall; and the others were identified as different birds by the presence of a colour ring, their age, variations in their plumage and/or their gender. Similarly, in the town of Aylesbury, Buckinghamshire, at least five different Peregrines were identified using a church in autumn 2010. Alongside the usual pair, the other individuals showed differences in plumage colour and age. And in the centre of Bristol various individuals have been spotted in winter, identified by their habits, plumage, sex and rings each year. In Bristol in 1999, two Peregrines were frequently seen perching on an office block in the city centre at any one time. It was easy to assume these were the only two falcons using the building, especially as Peregrines were still unusual in cities in England. However, one day a male Peregrine hit the building and was found dead below. Almost immediately, two birds continued to use the concrete structure, revealing that more than two had been in the area and using the building. It helped prove further that at any one time there are likely to be a number of Peregrines in or over a town or city.

Individual markings also allow Peregrine researchers to work out who is breeding with whom. In Germany and Sweden extensive colour-ringing studies have meant the majority of Peregrines are marked, and in the process remarkable exchanges between individual birds have been revealed. For example, Peregrines tend to have a number of partners during their lifetime, rather than just one. Female Peregrines may have between one and five different partners while males may have one to three. During the Second World War Peregrines were often shot to stop them from killing pigeons flying over with important messages. One female Peregrine managed to pair with three different males in the space of a month as each was shot during that period.

In the North
For Peregrines living in colder, more northern climes, winter is time for migrating and moving south. Many thousands of Peregrines depart from North America, northern parts of Europe and Siberia to winter in countries where prey species of bird have also flown.

In northern Europe and the Arctic Circle, Peregrines nest on the ground. Across the tundra, pristine raised bogs and aapa mires or patterned fens are full of plants dominated by *Sphagnum* species, and their raised, drier vegetated areas with dwarf shrubs provide suitable nesting habitat for Peregrines. Where trees exist, Peregrines will use old nests of eagles and Ravens *Corvus corax*. During the summer, there is an abundance of daylight and food – breeding wading birds, finches, buntings and waterfowl. However, as the days draw in, most of the birdlife departs and there is little or no sunlight. The Peregrines' only choice to survive is to move south. Those living in the very north, in Lapland, will tend to migrate further south towards northern

Africa and southern Spain, while those living in southern Scandinavia will only go as far south as France and northern Spain. Many also make it west into the UK, and often spend time in towns and cities there. This is a good example of 'leapfrog' migration and it is found in many groups of birds, such as wading birds. Ringed Plovers *Charadrius hiaticula*, for example, almost replicate the movements shown by the different Peregrine populations. Migrating Peregrines may move between 126 and 196 km (78 and 122 miles) per day, and like many raptors will complete their journeys mainly by gliding rather than flapping. When Peregrines are seen drifting over occupied territories, they are often very high and gliding through the air. These are thought to be satellite birds, non-breeders and nomads, moving around the countryside looking for vacant territories and becoming familiar with the terrain. If a territorial pair spots such a bird, they will hasten to see it off.

In the UK, increased sightings of Peregrines off the east coast in the autumn may be migrants heading south from Scandinavia, and ringing recoveries are certainly helping to prove this. In 2009, a juvenile Peregrine was found injured on the Somerset Levels in the west of England and taken into care – it had a Swedish ring and was traced back to a nest site in southern Sweden.

Preparing to breed

For those Peregrines breeding in cities and towns, winter is a period when they stay close to the nest site (Fig. 4.3). In mid-winter they may visit the nest itself, perhaps even sitting in the scrape and appearing broody. They will also use other nearby buildings but will be less obvious as they spend much time roosting or away catching prey. They will also store provisions during cold weather, stocking up on birds such as Woodcock, Teal *Anas crecca* and Moorhens *Gallinula chloropus*, and keeping them in a cache, the equivalent of a fridge freezer!

Figure 4.3 On cold winter days Peregrines keep watch from their roosts, conserving energy, and often stay in the same place for much of the day (Matt Allen).

A Year in the Life of an Urban Peregrine 53

Often this may simply be a gargoyle or the flat roof of a building (Fig. 4.4). Strong winds often blow such items down to the roads and paths below, but they won't be retrieved by the Peregrines if this happens.

As winter draws to a close, Peregrines will be busy strengthening their pair bond – often through calling, postures and spectacular flight displays. On a nice sunny day in February, Peregrines can be seen diving at vast speeds across their urban landscape before pulling out of the flight to impress a mate.

Even before any eggs are laid, copulations may begin as early as the end of December and continue through January, February and March. This is also a time when other Peregrines passing over will get chased off. If they don't get the hint, a fight may break out with physical contact, and in some instances Peregrines may even be killed during such an intruder interaction. This often occurs where the falcon population density is high and a greater number of intruders are passing through established territories. Historical records show instances of this happening in Montreal and Mallorca, and in both instances the dead individual had been partially eaten. Perhaps once the rival is dead, the other Peregrine treats it as prey. In 2013, a fresh dead male Peregrine found below an urban nest site in Derbyshire was also thought to have been killed during a fight, as two live birds remained on the territory following his death. Interestingly enough, another Peregrine was found dead at the same site in 2012. It appeared to have died of natural causes, probably through trauma during a fight. In rural locations, dead Peregrines have also been found eaten, being likely victims of earlier territorial disputes (Fig. 4.5). Peter Walsh, who has been studying Peregrines in Cornwall since the mid-1980s, observed at least two intruding males killed by a pair living on the coast during one season.

Figure 4.4 A Peregrine's cache is the equivalent of a fridge freezer with a variety of previously killed birds ready to be eaten on a lean day. Here lies a Woodcock (top) and a Song Thrush (bottom) (Dave Pearce).

Figure 4.8 A pair of Peregrines mating – the 'kissing' of their cloacas is over within seconds (Matt Allen).

male regress into tiny organs, and they only develop again during late winter when hormones trigger their growth. This reduction in size helps to keep the birds light and avoids them carrying excess weight when it isn't needed.

Before a fertile egg is laid, the cells within it will begin to divide and develop. As the balls of cells make their way through the reproductive tract of the female they get embedded in protein-rich yellow yolk and then enveloped in albumen, the white part of the egg. Prior to laying, special cells in the tract deposit an even layer of plain-coloured shell, and further along glands deposit the rich brick-red pigments typical of Peregrine eggs. During this time the female will be in need of calcium carbonate to produce the eggshells – this may come from the bird's own body as well as her diet.

The very early development of the egg will remain suspended until it is incubated as part of a full clutch. During this time, the breast skin of both the male and female Peregrines becomes thickened and engorged with blood to provide a perfect surface area to lay over the eggs and keep them at a constant temperature during incubation. The feathers of birds grow from distinctive tracts – the breast and abdominal feathers grow along tracts on either side of the breast and are long enough to spread across it. However, tiny, fluffy body feathers may cover the breast skin itself and during incubation will be moulted out.

Early on the morning of laying the female will become fidgety and finally lay her egg onto a thin layer of substrate. Over the next 48 hours the next egg

Figure 4.9 A clutch of four eggs may take just over a week to lay (Dave Pearce).

will go through the same process, and it may take over a week for a clutch of 4 or 5 eggs to be laid. The full clutch will vary between one and five eggs, although six have been recorded in North America on a few occasions. Three or four eggs are most usual (Fig. 4.9), and data from the BTO suggests that Peregrines are laying significantly smaller clutches long term. However, in Bristol the Avon Gorge pair laid five eggs in 2008, 2010 and 2011, and at a few other UK sites, such as Sutton and Wrexham, Peregrines have also laid a clutch of five.

During cold, wet or windy weather, and at night, the pair may begin covering the eggs before the full clutch is laid. In very cold weather the eggs may be covered for long durations and this may even initiate the development of the eggs early. During March 2013, urban Peregrines continued to lay eggs during sub-zero temperatures and heavy snowfall. Adults were observed incubating the first eggs, or at least stopping them from freezing. Some nests were completely covered in snow just prior to egg laying – and in Derby a warm hot water bottle lowered down to the nest platform meant the birds had a snow-free scrape. Many pairs were incubating their eggs to the maximum incubation times, suggesting the eggs were taking longer to develop during this colder spring (Fig. 4.10).

However, in more usual, milder springs the parents will begin incubating the eggs once the final egg has been laid. The use of web cameras at urban sites means that many urban nests across the UK can be monitored during the egg-laying period. This provides a chance to find out when eggs are laid, the size of clutches, and hatching dates, generating important nest record data which is used by the BTO. This type of information can be used to study the productivity of bird species, changes in egg-laying dates, and reasons behind less successful nesting attempts or declines in a species' population. Such data collection can reveal, for example, that in Australia, earlier-nesting female Peregrines tend to be more successful at breeding and show a bias towards producing female chicks.

Both the male and female Peregrine will incubate the eggs, though the female will do a greater share and covers the night shift. Males may incubate

Figure 4.10 The incubation period is a relatively quiet time for Peregrines as they wait for their eggs to hatch (Dave Pearce).

anywhere between a third to half of the time in some cases. The eggs hatch roughly 4 to 5 weeks after incubation has begun, taking anywhere between 28 and 36 days. In the UK this falls towards the end of April and the beginning of May, and during colder weather clutches may hatch towards the end of this period. If a pair has begun incubation when their full clutch is complete, then the chicks all hatch within a 48-hour period. Often all the eggs within a clutch will hatch, but sometimes one or even two may not. These are infertile or the embryo may have died during incubation. Weak or particularly small chicks may also die within a day or two of hatching. Generally, poor weather, particularly in April, can lead to smaller brood sizes. The weather, experience of the parents, and the presence or absence of an inexperienced helper may all play a part in chick survival. Younger birds who are less experienced with incubating eggs and clumsier in how they cover and turn the eggs may also find their clutch hatching later.

If a Peregrine fails to hatch its first clutch through predation or desertion, a new clutch may be laid, especially if the first set of eggs was lost close to the beginning of the incubation period. A new clutch may be in place around three weeks later, and will result in young Peregrines still being reared in the nest into July.

As soon as incubation begins the balls of egg cells within each egg start to develop again, and within just a few days the embryo will begin to resemble the shape and form of a chick. The heart develops quickly and begins to beat, and by a week old within the egg the chick will have a proper beak and feathers. In the final week before hatching the chick is well developed, and a few days before hatching it will begin cheeping and communicating with its parents. On web cameras the cheeping can often be heard and gives an indication that hatching will be soon. The neck muscles in the chick are well developed, and using the egg tooth, a hard, pointed area on the top mandible, the chick taps and pushes against the eggshell until it cracks and begins to open up. The chick will be wet, weak and exhausted, but over the next few hours it dries out and becomes a fluffy chick ready to be tended to and fed

A Year in the Life of an Urban Peregrine

by its parents. The eggshell itself will usually remain in the nest and be trampled over time by the parents and chicks (Fig. 4.11).

As the eggs are close to hatching there is also a change in the behaviour of the incubating parents. They are restless, may call quietly back to the chicks, and when the chicks have just hatched the adults characteristically raise themselves up slightly, with their wings dropped down. This is a change from the more relaxed incubating bird, sitting down comfortably with its wings in a natural closed position. The parents may also become more aggressive and territorial at this time.

Figure 4.11 When the chicks hatch the parents move the eggshells to another part of the nest rather than eating or removing them completely. The chicks may all hatch over a period of a day or two (Dave Pearce).

As with a human baby, each day shows sometimes subtle, sometimes obvious changes in the development of a Peregrine chick. During the first week after hatching the chicks line themselves up, huddled together ready to be fed by their parents (Fig. 4.12). With remarkable care, the adults will take tiny pieces of food and delicately feed them to each chick. It isn't long before a hierarchy may develop and some chicks may demand or be strong enough to receive food first, while weaker, smaller chicks get the food last. In lean times, these younger, smaller birds may receive little or no food and die, but in times of plenty they will survive.

For the first week or two, the chicks will be covered and kept warm (Fig. 4.13). During these first two weeks the young still have their first coat of down which dried out upon hatching, but the small size of the chicks means they have a large surface area to volume ratio and can lose heat and energy very quickly. Generally it is the female who keeps them brooded constantly for the first eight to ten days, but this may extend to two weeks or more if the

Figure 4.12 Peregrine chicks only a few days old (Dave Pearce).

Figure 4.13 The chicks' parents cover the young for the first few weeks of their lives. Here the female is carefully curling her talons to avoid stabbing the young by accident (Dave Pearce).

Figure 4.14 The young Peregrines line up and take it in turns to be fed. The female chicks are already bigger than their male siblings (Dave Pearce).

Figure 4.15 At just over three weeks old, the wing feathers of the young Peregrines appear and they become more active around their nest (Dave Pearce).

weather is wet and windy. During this time the male is doing most of the hunting and bringing back prey, although sometimes the female may leave and hunt too. Whether this is related to lack of food supply or how good her partner is at hunting is not known. Either way, at a very young age hardly a moment will go by when one of the parents is not sitting over the chicks, keeping them warm and sheltered. Food will be offered as tiny titbits, delicately offered to the helpless chicks which are hardly able to lift their heads. It is remarkable how such a tiny chick is able to feed at all. But as each day passes they gain more strength and increase in size, and after a few weeks they will have grown a second layer of down which keeps them even warmer. Depending on the weather, it is around this point that the chicks will be left by themselves for longer and longer periods. However, during poor weather such as rain and wind, a bedraggled parent can be spotted standing over his or her dry, fluffy white young to protect them from the worst of the elements. Any covering of the chicks ceases after about three weeks. By this age I am always surprised at just how big they have grown, especially the young females, which are always larger than their male siblings. During their development the chicks also become very vocal, and they are often heard before they are seen! The chicks continue to grow and put on weight – so much so that by the time they fledge they will weigh more than their parents (Fig. 4.14).

Figure 4.16 An adult feeds one of its young just after it has fledged the nest (Hamish R. Smith).

Their contour feathers, the main wing, tail and body feathers, will also be growing through by this stage. The feathers start off as light blue pins which are rich in blood vessels. Inside, a pulpy soup of cells is organising itself into regular lines of barbs and barbules which form the zip-like processes of the feather. Gradually, feather tufts emerge from the pins and the feathers begin to grow out, uncurling and straightening as they do so. The second coat of down becomes dotted with darker body feathers, and the head in particular begins to develop the juvenile colours (Fig. 4.15).

Once the chicks are a few weeks old the parents will leave them for significant periods of time, and each parent will go off hunting. David Gittens, a volunteer for the Hawk and Owl Trust, has diligently observed via web cameras the feeds given to urban Peregrine chicks on an artificial nest platform at Norwich Cathedral in Norfolk. Feeds varied between two and eight per day, with an average of five, and the first feeds of the day were generally between 4.30 am and 7.45 am, apart from a particularly rainy day when the chicks weren't fed until early afternoon, although they were quite big by this stage. The last feeds were around 8.30 to 9 pm. David found that during the first few days of the chicks' lives they were fed only every few hours – though at this stage the young would still be living on their own yolk sac, which would be present in their abdomens before being reabsorbed. Feeding rates quickly increased to between six and eight per day, before dropping off to between four and six per day, and intervals between feeds increased from two to four hours (Fig. 4.16). Towards the end of the chicks' development less food was brought in as the parents were encouraging them to fledge. Studies of Peregrines on a wider scale have shown that feeds may vary between 4 and 11 per day, and the intervals between them can be as little as 10 minutes or as long as 6 hours! Generally, feeding rates will vary depending on the size of the brood.

As the chicks develop their feathers continue to grow, covering the young birds in their juvenile plumage which will allow them to fly and give them protection from the weather over their first year or two of life, before they moult into their adult plumage. Even at an early age, the young birds will begin to flap their wings and strengthen their wing muscles (Fig. 4.17). Stretching, nibbling and flapping all become regular parts of their daily routine, along with sleeping, dozing and following flies, gulls and planes overhead (Fig. 4.18). At the time of fledging, around five weeks after hatching, the wing feathers of the chicks still won't be fully grown and they will still have tufts of down feathers around their head, back and tail. The egg tooth will often remain on the beak of the young bird all the way through to fledging. The chicks often begin flying with feathers still in pin and they are also heavy at this stage, with large female chicks sometimes weighing up to a kilogram, if not more. Flying can therefore be clumsy and often lead to the chicks becoming grounded. Inquisitiveness, wind, fellow siblings pushing and shoving, slipping and hunger are all reasons

Figure 4.17 Wing flapping takes up a lot of a young bird's time as it strengthens its wing muscles (Dave Pearce).

Figure 4.18 As well as resting between feeds, preening (right) is an essential part of the chicks' daily routine (Dave Pearce).

A Year in the Life of an Urban Peregrine

why a young bird may inevitably fledge – it has got to do it at some point, but how it does it can vary a lot.

On fledging in urban areas the chicks frequently end up grounded and need a few days to strengthen their wings and possibly lose some weight! Many urban nest sites are close to rivers and other waterways, and the steep-sided brick or concrete banks offer little comfort for young Peregrines which may fall down into the water or be mobbed into it by urban gulls. If they don't receive help from nearby people they may die from drowning or hyperthermia. More gentle, grassy banks allow young Peregrines to swim to the edges using their wings and then climb out, dry off and recover (Fig. 4.19). This is probably a natural adaptation to historically living on cliffs along the coastline – chicks may sometimes fledge inadvertently into the sea. Interestingly, a female coastal Peregrine in Cornwall has been observed recovering a chick from the sea, bringing it back to the safety of the nest. The chick survived with no ill effect.

Figure 4.19 A young Peregrine fledges into the water and swims to the edge to dry off (Des Bowring).

Once the chicks are out of the nest they will perch on the building they were hatched on, or on nearby roofs, still sporting their downy head feathers. The chicks can usually be heard before they are seen, screaming loudly all day long, begging for food (Fig. 4.20). When the parents do bring in food, the young will fly up to them and feed themselves (Fig. 4.21). This is a very exciting time to be watching Peregrines – alongside the parents, up to four or five young can all be flying in the sky together. And once the young birds fledge, there is no stopping their antics (Fig. 4.22). From short, fast stoops to chasing each other or gulls, the juveniles are strengthening their flight muscles, honing their flying skills and learning how to hunt for themselves (Fig. 4.23). As the parents bring in prey, they soon catch the attention of their offspring and spectacularly transfer the food in mid-air.

Figure 4.20 From a good vantage point a juvenile Peregrine begs for food (Sam Hobson).

Figure 4.21 Two siblings squabble over food mid-air. The larger female is on the left (Sam Hobson).

Figure 4.22 A young bird (left) turns upside down to try and grab a pigeon from its parent (Sam Hobson).

Over the next few weeks the parents will begin to deliver live birds, usually pigeons, and release them for the chicks to catch. At first there are plenty of near misses, but gradually the chicks learn to successfully secure the dropped prey before it falls to the ground, often amongst busy shoppers in some cities! As the summer goes by the young birds become sleek, fit and fast, and begin catching their own birds and venturing further afield (Fig. 4.24).

As July turns into August, sites with no breeding Peregrines may see adult birds return after an absence since April. Meanwhile, juvenile birds will hang around with their parents and slowly

Figure 4.23 A young Peregrine feeds for itself, in this case on a Swift brought in by one of its parents (Dave Pearce).

Figure 4.24 A juvenile Peregrine succeeds in catching its own prey (Sam Hobson).

territory holders no doubt chase them off as they pass overhead. Juvenile birds won't usually begin nesting until they are a few years old, although in an ever-increasing population which is doing well young birds will often attempt to nest in their first year, though often unsuccessfully or in a haphazard way. However, with nowhere available to nest the young birds may decide to stay with their parents.

In Japan, where this has often been observed, it is thought to be typical of a non-migratory population where prolonged juvenile-dependency periods are more common. Peregrines are long-lived birds and if their mortality rate is low in urban areas, with adults staying in their territories for many years, then there is a relatively low dispersal rate of adult birds. This leaves little scope for young birds to find new territories and therefore some young birds delay their own dispersal, leading to cooperative breeding. The opposite may be true for rural Peregrines which are persecuted, for example in the uplands of the UK. Here territories are annually becoming vacant as birds are shot, trapped or poisoned, so young birds being reared at nest sites in both rural and urban locations nearby are able to disperse and find new territories where they can feed or attempt to nest. If left alone they may be successful at nesting – but more often they will also be persecuted.

As Peregrines increase we are likely to see cooperative breeding happening more often. Peregrines will continue to nest in closer proximity, with distances between nest sites reducing to less than a kilometre in urban areas. In Germany, two pairs of Peregrines have been recorded nesting 300 m apart at a power station, each facing in an opposite direction. In the UK there are already increased interactions between breeding and non-breeding Peregrines, as well as new breeding pairs squeezing into available but possibly inferior territories. In 2013 we had confirmation that two pairs of Peregrines were nesting very close to each other in Bristol, with a new pair setting up a nest only a kilometre from an established nest site of 23 years in the Avon Gorge. Of the new pair, the female was a brown one-year-old while her mate was an adult male. They laid two eggs and incubation commenced. However, she then stopped visiting the nest while the male continued to incubate on and off for a week or two. Around the Bristol area, occupied Peregrine nests are more usually 3 km apart from each other. During the 1990s, Derek Ratcliffe wrote that the distance between Peregrine nest sites in some places was reducing as their population was increasing. Nest sites were more frequently being found less than 0.5 km away from each other. When Peregrines were more common on the northern moors in the UK, their

Figure 4.28 A pair of Peregrines copulating – but is the male or female sneaking off to mate with another bird? (Sam Hobson).

territories were also much closer together. As Peregrines continue to increase, no doubt we will see more territories appearing and the distances between them reducing. Over time as persecution reduces, we may see this happen again in rural Peregrines too.

Extra-pair copulations

Like many bird species, breeding Peregrines will sometimes disappear off to mate with another Peregrine whilst their partner is unaware (Fig. 4.28). This behaviour, known as extra-pair copulation, has been observed in many bird species, including the Great Tit, Blue Tit and Barn Swallow *Hirundo rustica*, as their behaviour, genetics and pairings have been studied in greater detail. The classic faithful, monogamous partnership in many bird species couldn't be further from the truth. Observations in Germany have revealed that female Peregrines often sneak away just prior to egg laying to mate with another male, and genetic tests on offspring have revealed that the young have not only been fathered by the resident breeding male. For the female Peregrine this is a great way of increasing the genetic variation in her offspring, perhaps with a better quality male, and ensuring the best genes are passed on whilst still having a resident partner to help her rear the chicks. But who's to say he hasn't been off to mate with another female? For him this is an ideal way of increasing the number of young he fathers in one year.

Peregrines don't exhibit this behaviour all of the time, and some studies on urban Peregrines show that many pairs do remain very faithful and stay together for many years at one location. However, in Sweden researchers have studied females who pair with three to four different males during their lifetime, and have found males who pair with two or three different females during theirs.

Polygyny

One behaviour to keep an eye out for in urban Peregrines is polygyny, where one male breeds with two separate females at two separate nest sites. While this isn't observed often in Peregrines, it does occur from time to time. For example, it has been recorded in Germany where the population is very well studied, and there is one instance in downtown Toronto, Canada, where two city-nesting females were sharing the same male only 0.6 km apart from each other. Polygyny is more widespread amongst other raptors such as harriers, hawks, and rodent-eating falcons such as Kestrels. Generally it is a behaviour found in raptors that are more reliant on voles, and in those species which have weaker territories and change partners from year to year. Peregrines, on the other hand, are bird hunters and tend to keep their territory from year to year.

Even if suspected, the behaviour can be difficult to confirm if the individual birds involved are not tagged or colour-ringed in some way. When we suspected polygyny was happening in Bristol with two nests very close to each other, we were able to confirm this wasn't the case by having observers at both sites at the same time and in constant communication. As one pair were at observed at one nest, it was clear a separate pair were at the other site.

More observations and research are required to understand why polygyny may occur in Peregrines. However, it might relate to males defending good quality territories with plenty of food and being superior to other nearby males. A second female may be willing to nest within the high-quality territory of such a male rather than pair up with a less superior male in a poorer quality territory. By doing so, she is more likely to raise chicks, and the winning male is able to increase the number of offspring he sires in that one breeding season.

Inbreeding
I first discovered inbreeding was taking place at one of my study sites when the male we had colour-ringed 'AA' as a chick appeared to be breeding with his mother (Fig. 4.29). With some DNA testing of their moulted tail feathers we confirmed with a high certainty that indeed the two birds were related. A mother–son relationship is the common pairing if incest occurs in Peregrines, whereas a father–daughter relationship has not been observed. Inbreeding has also been reported in Germany, in Belgium, and at a few locations in the USA. There are also a few examples of half-sisters and half-brothers pairing up, as well as full siblings which both share the same father and mother.

Peregrines do have a mechanism for avoiding inbreeding, with females dispersing away from the nest much further than males. And parents appear to be able to recognise their own offspring, as shown in cooperative breeding situations, so it is odd why incest should still occur. Any incest is unhealthy for a population and should be avoided at all costs. The resulting offspring are likely to be weaker, and see their survival and reproductive success compromised. While examples of incest in Peregrines have only been reported recently, this may be due to the information gleaned from colour-ringing and DNA testing. However, it is still a relatively rare event, although in the Midwestern USA, where Bud Tordoff and Pat Redig helped Peregrines

Figure 4.29 Peregrine colour-ringed AA (left) with his mother who he breeds with in Bath (Hamish R. Smith).

re-establish themselves, they discovered inbreeding occurring in 4 per cent of the 454 nesting attempts.

Why some Peregrines don't avoid breeding with their offspring or siblings is unknown, but may relate to changes in how young Peregrines disperse when populations are at a higher density, as well as to the mating system Peregrines employ. For example, although this is untested, a female Peregrine paired with her son may use extra-pair copulations with other males to avoid some incest, while still having a devoted son to help care for his offspring or half-siblings. Even if he hasn't sired the chicks, they are still his kin as they have his mother's genes. Meanwhile, due to an abundance of Peregrines he decides not to disperse far because there is nowhere else for him to set up home. Unlike when Peregrines were rare, there do appear to be lots of Peregrines present in areas where inbreeding has been recorded more recently. Therefore, it is less likely to be due to the absence of other individuals to breed with. In reality, the exploration of kin recognition, inbreeding and mate selection is complicated, and further studies on specific individuals may help reveal the truth behind why some Peregrines don't avoid breeding with their close relatives.

Chapter Five
Food and Feeding

While watching Pied Wagtails *Motacilla alba yarrellii* coming in to roost and feeding at a car park in the village of Charmouth, Dorset, I suddenly heard a 'whoosh' and looked up to see a Peregrine dashing past metres away. The wagtails fled, and the falcon carried on past chasing a particular wagtail it had singled out. The tiny bird dived and dodged the falcon and escaped, while the Peregrine flew off rather sheepishly. However, it was not to be beaten, and five minutes later I heard another 'whoosh' – before I could blink the Peregrine had returned, this time snatching a wagtail in its talons and flying off with it. It all happened so fast, and although the wagtails were keeping their eyes peeled for danger this one wasn't wary enough!

Studying what Peregrines eat is what got me hooked on them. Not only was I excited that these falcons were living in Bristol, the city I had come to be a student in, but also that they were eating birds I had never seen in the hand or up close before. And I hadn't appreciated just how varied their diet was. On the cloudy, overcast October afternoon in 1998 when John Tully took me to Broadmead in the centre of Bristol, I remember looking up, and on the tallest building in Bristol there was a Peregrine standing near the top. It wasn't doing a huge amount; it was just standing upright, grey and still. But nonetheless, a Peregrine it was. The building manager, Denzil, would pick up any dead birds the Peregrines had dropped and put them in a bin in the stores. Every week, John would drop by and see what was there. On the day I went with John there were the bright yellow-spotted feathers from a Golden Plover *Pluvialis apricaria*, something I had never seen so close before. Additionally, a half-eaten Lapwing *Vanellus vanellus* was lying there, along with various wings and heads of Redwings *Turdus iliacus* and Fieldfares. From that day on I was hooked. While it may seem morbid, Peregrines are

Figure 5.1 Catching other birds is vital for the Peregrine, especially if it is to rear three or four chicks in one season (Dave Pearce).

catching and eating these types of birds all the time. And rather than them being thrown away or swept up by the street cleaners, John was using them to glean some fascinating insights into the diet of urban Peregrines. I was privileged that John handed this job over to me during this early period of my life with Peregrines. During this time I got used to picking up feathers and legs outside Pizza Hut while people ate inside – the restaurant happened to be right next to the building where the Peregrines dropped their food! Collecting prey isn't something you can explain easily to passers-by. I got used to explaining myself to those visiting the church in Bath as they checked I wasn't up to more sinister activities.

Peregrines are renowned for their speed and ability to catch prey, mainly birds. Over millions of years the Peregrine has evolved to be an efficient predator, feeding on the flesh of other birds which it catches in flight. Its hooked bill and sharp talons provide the perfect tools for the falcon to effectively catch its prey, the bill having evolved quickly as the species competed against other birds. The Peregrine needed to be fast to take prey other species were unable to catch, pluck and eat themselves (Fig. 5.1).

Peregrines are superb carnivores, an apex predator at the top of their food chain. Although they feed almost entirely on birds, they sometimes also eat mammals, including rabbits, bats and squirrels, while fish and reptiles such as snakes and lizards have also been recorded. In some parts of the world, Peregrines themselves may get eaten by other raptors such as Great Horned Owls *Bubo virginianus*, and even in the UK Goshawks and Eagle Owls pose a threat to adult and young falcons.

What is their success rate?

The ability of a predator like a Peregrine to catch another animal is an incredible feat. Watching their speed, instinct and skills in making a kill is a marvel. However, those hunts that result in the Peregrine catching and killing its prey don't happen every time. There are many times when the bird being pursued has the upper hand and manages to escape. I recall watching a Peregrine chasing a pigeon above a Bristol shopping centre, and just as the

Figure 5.2 A Peregrine clutching a Feral Pigeon after a successful hunt (Sam Hobson).

falcon was about to close in the pigeon suddenly dropped out of the sky – an effective escape technique for the pigeon (Fig. 5.2). However, on another occasion I was looking out across the Avon Gorge when I heard a 'whoosh', and dashing past me a female Peregrine was gathering speed across the River Avon. Unbeknown to me, a pigeon was flying on the other side close to the top of a cliff. Before I knew it, the female had taken the pigeon by surprise and grabbed it. The attack was quick, opportunistic and full of stealth. For breeding Peregrines, up to a third or more of hunts will be successful and lead to a kill. For migrating falcons moving through a site, this may drop to only a tenth of all attempted hunts.

While Peregrines don't strike a bird every time, enough kills are made to survive and to have some birds left over to store as a surplus in a cache. Sometimes kills are made and the dead prey is simply left and not eaten. Dick Treleavan and Dick Dekker have both studied the hunting behaviour of Peregrines in huge amounts of detail, spending hours watching their antics. While many Peregrine attacks are for killing prey to survive, others appear to be 'play' dives that intentionally miss their prey – these false attacks still include the chase and dive, but individuals pull out before they would normally strike. Just like a well-fed cat, Peregrines sometimes catch a bird only to let it go again alive (Fig. 5.3). Perhaps this behaviour helps Peregrines to hone their hunting skills and maintain their muscle tone. When a Peregrine has eaten a pigeon or another large bird such as a duck, it may not need to eat again for a few days and may just spend the next few days resting on a favourite building. The occasional tussle with another bird may be a way of keeping it fit and trim. Despite these dummy hunts, when a Peregrine wants to kill there is something about its presence and attitude that shows it means business. Hunting attacks that lead to a kill tend to be full of energy, purposeful, and are followed through – there is no giving up or half-heartedness to it. A kill in this instance is usually made within three or four minutes.

In an urban environment there are a number of factors to consider which may reduce the likelihood of a successful kill. An urban environment is not the safest one for diving at up to 290 km per hour (180 miles per hour). Unless pursuits are performed high up, bridges, office blocks and other buildings can all pose hazards. In Italy, studies have shown that Peregrines generally hunt beyond a 1 km (0.6 mile) radius of where they usually hang out. Perhaps this distance away from their roost gives them the space to successfully hunt. Certainly in Bath the Peregrines usually disappear off to the outskirts of the city, returning later with prey and with little evidence of catching it close by.

Figure 5.3 This Common Gull fought back and survived, though with a featherless neck it may not have lived for long (Christine Raaschou-Nielsen).

It is common to see local pigeons, Blackbirds and Robins *Erithacus rubecula* at ease very close to a Peregrine roost – they are able to see the Peregrine, and the Peregrine has little chance of catching them at such close range (Fig. 5.4). Peregrines rely on surprise, speed and sometimes a chance encounter with a bird to be successful in catching and killing their prey. Peregrines will also

Figure 5.4 This pigeon is safe from being eaten. Birds within a few kilometres of an urban roost or nest site are generally not targeted during the daytime (Hamish R. Smith).

Figure 5.5 The Peregrine will reach at least 290 km per hour (180 miles per hour), if not more, during a stoop dive (Chris Jones).

hunt in pairs – this is thought to be a cooperative arrangement, with prey shared with a mate or offspring and not just eaten by the individual which makes the kill.

It is clear from examining the caches of urban Peregrines that they manage to kill and store tens or sometimes hundreds of birds, many of which may never get eaten. Peregrines catch a glut of prey items during a time of plenty, for periods when prey may be less available. For example, during freezing weather prey species such as larks, thrushes and ducks will suddenly disappear to warmer, snow-free places. With little prey left around (aside from pigeons), the Peregrines can make good use of their caches and feed on their larder. Observations also reveal that male Peregrines who have lost their partners can successfully rear chicks to fledging. It is suggested that both the caching of prey and good hunting skills make this feat possible.

The highest mortality for Peregrines is during their first year or two of life, and is often as a result of starvation. Up to half of the young that fledge will die in their first year – it is a huge challenge for Peregrines to successfully learn to catch prey over a long period of time, and those birds that are simply not able to kill enough food will starve. In the UK, this is a very similar mortality rate to other common birds such as the Blackbird and Robin, where 56 per cent and 41 per cent of first-year birds survive, respectively. Adult survival for Peregrines is much higher, at around 80 per cent, while for a Robin it may only be 42 per cent. So if a Peregrine can successfully hunt and find enough food, it has a good chance of surviving for many years.

How do Peregrines catch their prey?

Peregrines are fast birds (Fig. 5.5). When the Peregrine zoomed past me in Bristol in pursuit of a pigeon, it was probably travelling at over 161 km per hour (100 miles per hour). Swifts are certainly the fastest birds in level flight and can reach a remarkable 112 km per hour (70 miles per hour) in super-fast climbing flights and screaming chases. However, Peregrines are by far the quickest in a stoop dive, reaching over 290 km per hour (180 miles per hour).

And it is this stoop dive, the closing of the wings and dropping out of the air at immense speed to hit a bird, which is the most obvious and commonly recorded technique for the Peregrine to catch its prey. But it does have other tricks up its sleeve – techniques which work better in different situations. For example, Peregrines may take off from a perch and pursue a bird they have spotted in the air or close to the ground. Or they may fly low over a habitat and surprise a bird before it has time to escape. They may even go scrabbling around on the ground to find a bird that has dashed into some vegetation to escape.

Peregrines may kill their prey on impact following a stoop dive. However, often the victim is still wriggling or flapping and so the falcon simply bites the neck of the bird with its strong bill, using its tomial tooth to break the bird's neck and spinal cord (Fig. 5.6). Sometimes Peregrines will begin plucking their prey while it is still alive. The gradual compression and squeezing from the talons of the Peregrine, along with internal bleeding, shock and damage to tissue, eventually kills the prey. Birds are generally plucked and torn into pieces rather than being swallowed whole, and small prey may be eaten in mid-air (Fig. 5.7).

Figure 5.6 The tomial tooth on the upper bill of the Peregrine helps it to quickly kill its prey.

Figure 5.7 Peregrines will feed on the wing – here an adult is picking at some food clutched in its talons (Andy Thompson).

What do Peregrines eat?

Peregrines eat mainly birds. And while they do catch and eat other animals from time to time, especially small mammals such as bats and young rabbits, their technique of catching birds is their best attribute for survival. There is a common assumption that Peregrines just eat pigeons, but as I will explore below, while pigeons may be important in their diet they are by no means the only things they eat. Indeed, in an urban Peregrine's diet pigeons may only comprise half of everything they eat. Individual Peregrines will also favour certain birds which they catch on a more regular basis than other birds. For example, some may focus their efforts more frequently on gulls or pigeons, while others may target Swifts or Woodcock.

What are the common prey items?

In bags of weekly remains that I receive from other Peregrine researchers it is easy to spot the odd feathers out, such as those from wading birds, woodland birds and seabirds, amongst the pigeon feathers. I am never quite sure what I may find, and it is exciting to imagine what has been flying near

Figure 5.8 Feral Pigeons are very much at home in an urban environment, and provide Peregrines with plenty of food (Adam Rogers).

or over our towns and cities, and what the Peregrines have managed to intercept. The Peregrine's diet is an interesting talking point for those studying this bird, as well as for people who have just seen one for the very first time. It comes as a surprise to many that they don't just eat pigeons; it captivates people when they discover what else Peregrines eat and how they go about catching their prey. Most of the species I discuss below relate to the UK and Europe, but many of the same or equivalent species are also taken by Peregrines all around the world.

Pigeons

Pigeons, of the feral, street pigeon type, are by far the most common prey item, making up anywhere between 10 and 60 percent of the Peregrine's diet in urban locations, with the bulk taken during the breeding season (Fig. 5.8). The number of pigeons available can dictate the species' productivity – recent changes in flight routes of racing pigeons in the Welsh Valleys have resulted in a drop in the number of chicks reared by Peregrines. Where the variety of prey species increases and fewer pigeons are eaten, the average number of chicks reared by Peregrines generally goes down. Pigeons are common birds in all towns and cities, as well as in industrial locations. And while they are fast fliers and have various techniques to dodge Peregrines, inevitably many don't get away in time. Young pigeons are more vulnerable and get taken in large numbers, but older, experienced pigeons are still vulnerable. For the Peregrine, pigeons are a staple food, something that they have been eating for millions of years. Originally they would have eaten native Rock Doves, but as pigeons became domesticated, their ability to survive and live in many different habitats saw them spread. They are very common across the world, and feral birds living on our streets are often regarded as pests. Additionally, the wild and feral pigeons are joined by those which are bred for their speed, appearance and flight around the world. Millions of racing and display pigeons live in lofts and gardens in the UK and beyond, with huge numbers raced across the countryside every year.

Collared Doves, Woodpigeons *Columba palumbus* and occasionally Turtle Doves *Streptopelia turtur* and Stock Doves *Columba oenas* also feature in the urban Peregrine's diet across the UK and Europe, but only account for less than a fifth of all the prey eaten each year. Collared Doves are the most common prey item out of these four species. Woodpigeons are taken in smaller numbers, perhaps a reflection of their bulk and speed (Fig. 5.9). Stock Doves are a more secretive woodland dove but do feed in open farmland areas where they are more vulnerable to being caught. The Turtle Doves are usually taken in very small numbers during the migration period and are often juvenile birds.

Figure 5.9 Woodpigeons are also eaten by Peregrines, but in much smaller numbers (Adam Rogers).

Starlings

Despite a decline in breeding Starlings in the UK and across Europe, they still feature highly in the diet of Peregrines – they are an obvious, gregarious flocking species which makes for an easy target (Fig. 5.10). During a typical year Starlings contribute to over a tenth of the Peregrine's diet, while during the breeding season 20 per cent of all prey may be Starlings. Their young fledge just at the time when a

Figure 5.10 A swirling, mesmerising flock of Starlings becomes even more spectacular when a Peregrine is on the hunt.

Figure 5.11 A Starling, like the Feral Pigeon, is accustomed to living off litter and food that people drop or leave behind.

Peregrine's own chicks are in need of quick and easy food. Such naïve and inexperienced birds make for easy pickings during May and June in the UK. Starlings are found throughout the world, and the species *Sturnus vulgaris* has been introduced to many countries, providing a semi-natural resource for Peregrines in countries from which the Starling doesn't originate. Millions of Starlings winter across European cities and provide a glut of food for Peregrines during the cold, short days (Fig. 5.11). With so many birds coming to roost, Peregrines are able to hunt the flocks, aiming for stray birds that may be on the edge or trailing behind. There may be many misses, but plenty more successes too. A Starling flock with a Peregrine in hot pursuit creates beautiful three-dimensional, globular shapes which move as waves in the air as every Starling is hurrying to get into the middle of the flock. In Australia, Starlings, which are an introduced alien species, have formed an important part of the diet of Peregrines. As Starlings have declined there, along with another species, the Silver Gull *Larus novaehollandiae*, researcher Jerry Olsen and his team have found that their appearance in the diet has also declined. Peregrines living on cliffs near Canberra have adjusted their diet and feed on other similar-sized native birds, alongside larger birds such as Gang-gang Cockatoos *Callocephalon fimbriatum*, Galahs *Cacatua roseicapilla*, Eastern Rosellas *Platycercus eximius* and Feral Pigeons. However, interestingly they have avoided other common species such as Sulphur-crested Cockatoos *Cacatua galerita*, Australian Magpies *Gymnorhina tibicen* and House Sparrows *Passer domesticus*.

Figure 5.12 Blackbirds are a common prey species for the Peregrine all year round (Gary Thoburn).

Figure 5.13 A Green Woodpecker is easily caught across an open landscape (Dave Pearce).

Woodland, garden and farmland birds

While Peregrines rarely visit gardens, they do feed on a variety of birds which we associate with gardens, parks and woodlands. House Sparrows, Great Tits, Blue Tits, Dunnocks *Prunella modularis*, thrushes, Blackbirds (Fig. 5.12), woodpeckers (Fig. 5.13), Treecreepers *Certhia familiaris*, Goldcrests *Regulus regulus*, Blackcaps *Sylvia atricapilla* and Chiffchaffs *Phylloscopus collybita* all feature in the diet. Blackbirds are taken throughout the year, but as soon as wintering Redwings (Fig. 5.14) and Fieldfares (Fig. 5.15) arrive from Scandinavia across western and southern parts of Europe they also quickly appear in the urban Peregrine's diet. Obvious, low-flying flocks of these thrushes make easy targets, and after a tiring night flight across the North or Baltic Sea, these birds, along with migrating Song Thrushes *Turdus philomelos*, Blackbirds and the less common Mistle Thrushes *Turdus viscivorus*, are easy for the Peregrine to catch. During the winter months, Redwings and

Figure 5.14 As Redwings arrive in the autumn they are quickly targeted by Peregrines (Allan Chard).

Figure 5.15 Peregrines easily spot Fieldfares as they roam across urban and country areas in big flocks (Gary Thoburn).

Figure 5.16 In the middle of a town or city it is common to find the feathers and other remains of Lapwings that have been eaten by a Peregrine (Gary Thoburn).

Fieldfares are the most likely prey remains to find alongside pigeons. Finches are also common prey items, especially Greenfinch, Chaffinch *Fringilla coelebs*, Goldfinch, and in some parts of Europe, Hawfinches *Coccothraustes coccothraustes*. Bramblings *Fringilla montifringilla* are usually taken in the winter along with Lesser Redpoll *Carduelis cabaret* and Siskin *Carduelis spinus*. Red-backed Shrikes *Lanius collurio* and Great Grey Shrikes *Lanius excubitor* appear from time to time, especially during their migration period, as do Northern Wheatears *Oenanthe oenanthe* and the occasional Eurasian Nightjar *Caprimulgus europaeus*, Bohemian Waxwing *Bombycilla garrulus* and Stonechat *Saxicola rubicola*. In countries such as Italy, Golden Orioles *Oriolus oriolus* also make an appearance in the diet. Quail *Coturnix coturnix* are regularly eaten during their spring and autumn migration periods.

Wading birds

While towns and cities may not seem the first place you would find a sandpiper or an Oystercatcher *Haematopus ostralegus*, millions of wading birds fly across the countryside and urban landscape every spring and autumn on migration to get to their summer or winter habitats. Wading birds normally associated with coastal estuaries and beaches appear in the diet of

Figure 5.17 Peregrines target flocks of Golden Plovers as they arrive in the UK for the winter or move across the country during cold winter spells (Gary Thoburn).

Figure 5.18 Whimbrel are regular prey items for urban Peregrines during the spring as they migrate over land (Gary Thoburn).

Peregrines at locations situated inland, at sites that may be tens or even hundreds of miles from the sea. Peregrines take the opportunity to feed on these birds as they pass over, often at night. Common wading birds found feeding on fields around the outskirts of towns and cities, such as Woodcocks, Lapwings (Fig. 5.16), Golden Plovers (Fig. 5.17) and snipes are eaten by urban Peregrines in large numbers. Most other species of wading bird also feature in their diet at some point, from Common Sandpiper *Actitis hypoleucos* to Whimbrel *Numenius phaeopus* (Fig. 5.18), Knot *Calidris canuta* to Avocet *Recurvirostra avosetta*, and Black-tailed Godwit *Limosa limosa* to Ruff *Philomachus pugnax*.

Seabirds

As with wading birds, many seabirds take an overland route to get to their breeding colonies. In the landlocked parts of Wales and Scotland, Peregrine researchers may occasionally discover the remains or rings of coastal birds such as Sandwich Terns *Sterna sandvicensis* and Kittiwakes eaten by Peregrines. And in Derby in the East Midlands, the city Peregrines took a Swedish-ringed Arctic Tern on a day when the only flock seen by birders was one passing north through England over a nearby reservoir. Terns (Fig. 5.19) frequently migrate over land and it is not unusual to discover the remains of Common, Arctic and occasionally Black Terns *Chlidonias niger*, while Little Terns *Sterna albifrons*, Sandwich Terns and Roseate

Figure 5.19 Terns, especially juveniles, are a regular prey item for urban Peregrines as they migrate over land during the spring and autumn.

Figure 5.20 Millions of Black-headed Gulls winter in towns and cities across Europe, and provide on ongoing food supply for urban Peregrines.

Terns *Sterna dougallii* may appear at urban locations closer to the sea. During rough autumn and winter weather, weak and storm-blown seabirds such as Storm Petrels *Hydrobates pelagicus*, Leach's Storm Petrels *Oceanodroma leucorhoa*, Manx Shearwaters *Puffinus puffinus* and Little Auks *Alle alle* stand out like a sore thumb as they flutter or stumble through the skies, and are easily picked off by Peregrines. You soon know when you have the remains of a petrel, as their distinctive musky smell lingers.

Gulls

Many populations of gull species are not considered seabirds anymore – those living in the heart of towns and cities in the UK have been thriving over the past 30 years, and while they may still visit the coast while on migration or during the winter, they are largely urban birds too. Black-headed Gulls *Chroicocephalus ridibundus* are the most likely prey item for Peregrines in towns and cities (Fig. 5.20). They are common, medium-sized, gregarious and relatively easy to catch, although a Peregrine may have a hoard of other gulls in its wake while going in for the kill. During the winter months, millions of Black-headed Gulls leave northern and eastern Europe and winter further west. In the UK, ponds, rivers, reservoirs and coastline are home to individuals of this species coming from Russia, Estonia, Sweden, Latvia and many more countries further north and east.

Peregrines generally leave the large gulls such as Herring and Lesser Black-backed Gulls alone and they rarely appear in the diet. They are big, successful, tough birds, and will fight back with strong bills and wings. However, Peregrines will defend their territories from bigger gulls and quite often the gulls end up injured where a Peregrine has hit them hard, breaking a wing on impact. The same may be replicated in other large birds in rural and coastal areas, including Gannets *Morus bassanus* and Ravens.

Corvids

I am always surprised that corvids such as crows are not eaten more often by Peregrines. Perhaps they are simply not easy to catch; they are very clever birds, after all. Carrion Crows *Corvus corone*, Rooks *Corvus frugilegus*, Magpies *Pica pica*, Jackdaws *Corvus monedula* and Jays *Garrulus glandarius* are all eaten by Peregrines, but interestingly most appear in the diet during the breeding season. Urban Peregrines tend to catch mainly Jackdaws, some Magpies and the odd Jay, usually individuals that have recently fledged and have wing and tail feathers still in pin (Fig. 5.21). Why Peregrines don't eat more of them during the non-breeding months is a mystery – perhaps they don't taste too good to the adults, whereas the hungry young Peregrines are less fussy about what they are fed. Corvids are very good at spotting predators such as Peregrines and are often first on the scene if one is present. They will use alarm calls to tell other birds that a predator is in their midst, they perch very obviously nearby, and they often mob the falcon while it is in flight. With such noisy neighbours advertising their presence, it makes it difficult for Peregrines to stand a chance of ever being able to catch them!

Figure 5.21 The wing remains of a Jay, a species usually taken by Peregrines during their breeding season (Hamish R. Smith).

Other raptors

From time to time Peregrines will eat other raptors. During 15 years of studying their diet I've only known a few Sparrowhawks, a Kestrel and half a dozen Little Owls *Athene noctua* to be eaten by Peregrines, most by the same Peregrine pair over that period (Fig. 5.22). Occasionally other species of owl such as Barn Owl *Tyto alba* may also be taken. Peregrines often live side by side with Kestrels and Sparrowhawks in quarries and close to woodlands, and don't

Figure 5.22 Little Owls are eaten occasionally by Peregrines.

Figure 5.23 The distinctive green feathers of a parakeet in the regurgitated pellet from a Peregrine (Nathalie Mahieu).

appear to be interested in catching and eating them. With equally keen eyesight and an awareness of what's around them, perhaps these other raptors just keep out of the Peregrines' way. Where they do nest close together, the different raptor species may in fact benefit one another. With more eyes on the lookout for a threat such as a fox, owl or marten, they can react together to see off the threat. Buzzards are often mobbed by Peregrines and sometimes killed, but not usually eaten.

Caged and escaped birds

Budgerigars *Melopsittacus undulatus*, Cockatiels *Nymphicus hollandicus* and other caged birds such as African Grey Parrots *Psittacus erithacus* are found from time to time in the diet of Peregrines, as they are in those of the Sparrowhawk and Goshawk. Colourful, naïve, and perhaps weak and lost, they are easily picked off by the Peregrines. Additionally, the increase in feral populations of Ring-necked Parakeets *Psittacula krameri* across the south-east of England, Amsterdam, Brussels and other cities across Europe provides a relatively recent and common source of food for Peregrines. At urban sites across London and the Home Counties, this species is found frequently in the prey remains and pellets of urban Peregrines, and at some sites dominates the diet alongside pigeons and Starlings (Fig. 5.23).

River birds and waterbirds

Despite their speed, whirring wings and shy nature, Kingfishers *Alcedo atthis* and Dippers *Cinclus cinclus* still find their way into the diet of urban Peregrines. The falcon's speed and agility mean even these fast-flying river birds can be caught. Dippers will often fly above the height of river trees such as Alder *Alnus glutinosa*, while Kingfishers fly out in the open across floodplains and lakes in towns and cities (Fig. 5.24). Ducks are also eaten in large numbers. Teal are by far the most common duck brought back to urban sites by Peregrines (Fig. 5.25). They are dinky ducks, and easy for Peregrines to grasp and carry. They are also a common species, particularly during the

Food and Feeding 91

Figure 5.24 Kingfishers are regularly taken by Peregrines in small numbers. Peregrines approach from behind and catch up with them very quickly over rivers and open areas (Pete Blanchard).

Figure 5.25 Teal are a very common prey species for urban Peregrines, and are easily carried back to buildings in the centre of towns and cities (Gary Thoburn).

Figure 5.26 Little Grebes also feature frequently in the diet of urban Peregrines, particularly during the migration period (Allan Chard).

winter, and will often be lurking in pools, ditches and lakes close to towns and cities. Despite ducks being fast, Peregrines are adept at surprising them by flying low over ditches and catching one as a small group rise upwards to escape. Other ducks such as Mallard, Gadwall *Anas strepera*, Shoveler *Anas clypeata*, Wigeon *Anas penelope*, Tufted Duck *Aythya fuligula* and Ruddy Duck *Oxyura jamaicensis* may also be brought back to urban sites, but in low numbers. Their heavy bulk means they are difficult to carry back to an urban site, and they are more likely to be eaten where they have been caught and killed. Little Grebes *Tachybaptus ruficollis*, Black-necked Grebes *Podiceps nigricollis* and Red-necked Grebes *Podiceps grisegena* are also eaten, especially during their migration periods (Fig. 5.26). Secretive crakes, rails and Moorhens feature highly in the diet of urban Peregrines – their appearance in the diet is explained in more detail later in this chapter. Occasionally herons and egrets may also be taken – some, such as the Grey Heron *Ardea cinerea*, Little Egret *Egretta garzetta* and Great Bittern *Botaurus stellaris* may be eaten where they have been killed or only taken a short distance away, while the tiny Little Bitterns *Ixobrychus minutus* in Europe may be brought back to the nest site.

Hirundines and swifts

Swifts are fast birds and when chasing each other in their evening swarming flights they may reach speeds up to 112 km per hour (70 miles per hour). However, the Peregrine can reach even faster speeds and the two aerial predators (the swift feeding on aerial invertebrates) accelerate through the skies with the Swift becoming the victim during the summer months (Fig. 5.27). While Swifts are not an uncommon prey item in the UK, they are not taken in huge numbers: fewer than 10 per year at any one site and making up just 9 per cent of the summer diet. This is perhaps a reflection of their ability to dodge the talons of Peregrines. However, in southern Mediterranean countries such as Italy where Swifts are more numerous, Peregrines are able to increase their success rates and Swifts may comprise 14 per cent of the summer diet. The remains of Swallows and martins, on the other hand, are only found in small numbers at urban sites in the UK, with perhaps just the odd one or two per year per site, if any.

Figure 5.27 Swift being delivered to two hungry juvenile Peregrines. Despite being quick, Swifts are still outmanoeuvred at times (Dave Pearce).

Bats

Peregrines will from time to time eat small mammals, from young rabbits to squirrels. Generally they form a very tiny percentage of their diet, and are rarely found in the diet of urban Peregrines. However, bats are recorded more regularly. They are still uncommon in the Peregrine's

Figure 5.28 Noctule bats are often eaten by Peregrines – these have been taken by Peregrines as the bats have been on migration (Peter Wegner).

diet in the UK, but are recorded more often elsewhere. For example, in Germany Peregrines will take advantage of the glut of Noctule bats *Nyctalus noctula* in the autumn as they migrate south to escape the cold (Fig. 5.28). In the UK, Peregrines have been found to take Noctule and Pipistrelle bats, usually catching them at dusk or dawn.

Ringed birds

The rings or unique identification markers of birds eaten by Peregrines are often found in their prey remains. Often the rings are still on the leg of the dead bird, but they are also found in the Peregrine's pellets or simply loose on the ground. Birds that have been ringed offer the opportunity to find out much more about where they have come from and how long they have lived for. I remember finding the silver ring of a bird on the University of Bristol's Wills Memorial Building with John Tully. Like most bird rings, it had a long number inscribed on it, but unlike most I am familiar with seeing, this one didn't have the address for the British Museum but instead one from Lithuania! We sent the details off and found out it was from a Black-headed Gull – no doubt one which had visited Bristol for the winter. Many other ring recoveries from Black-headed Gulls eaten by Peregrines come from Poland and other Eastern European countries.

In early May 2000, I visited a pair of urban Peregrines in Exeter with Peregrine researcher Nick Dixon. There were the usual remains of Feral Pigeons, Swift and Blackbird, but some other feathers and wings stood out. They were from a tern – but which species? With some more searching we found the bird's distinctive legs – tiny, short, orange webbed feet with sharp, hooked claws (Fig. 5.29). These can easily draw blood when terns attack people at their breeding sites. Excitingly, each leg had a silver ring on it – clues that would help us find out exactly which species the bird was. Once the details were sent off we found out the bird was a Roseate Tern, and had been ringed three years earlier at one of the largest breeding colonies of this species at Rockabill, a group of two islands near County Dublin in the Republic of Ireland. Before its death, the bird would have travelled to southern

Figure 5.29 The legs and rings of the Roseate Tern eaten by Peregrines in Exeter.

parts of Africa three times, and it would have been on its way back northwest to its breeding grounds.

There are numerous other case studies of ringed birds being discovered at the feeding and nest sites of urban Peregrines, and some sites may produce more rings than others. In the Bristol and Bath region I very rarely find any rings of birds, while in West and East Sussex, where many waterbirds such as waders and ducks pass through and winter on the coastline, Graham Roberts, who studies the urban Peregrines there, finds rings on a frequent basis.

If rings are found they can be reported online through Euring, www.ring.ac.

Nocturnal hunting

Figure 5.30 As Corncrakes migrate to and from parts of Africa they fly out in the open at night. As they move over urban areas they are easily picked up by Peregrines (Gary Thoburn).

Figure 5.31 Snipes are taken both during the day and night by Peregrines (Pete Blanchard).

As I began to study the Peregrine's diet in more detail I began to realise that some species were appearing in the ever-increasing list which I would least expect to find there. And it wasn't just in the UK – I was reading about similar, related species appearing in the Peregrine's diet around the world. In 2008, Nick Dixon and I had a paper published in the monthly journal *British Birds* summarising our research and reviewing what had been discovered across Europe and elsewhere. The appearance of more 'rural' or rarely seen birds in the diet of urban Peregrines in towns or cities around the world, including woodcocks, rails, plovers, flickers and cuckoos, was becoming commonplace. In the UK and parts of Europe, the presence of Woodcock, Little Grebe, Black-necked Grebe, Red-necked Grebe, Quail, Water Rail *Rallus aquaticus*, Moorhen, Coot *Fulica atra*, Corncrake *Crex crex* (Fig. 5.30), Spotted Crake *Porzana porzana*, Jack

Figure 5.32 Redwings commonly migrate at night and can often be heard on quiet, still evenings. As they migrate over towns and cities, they are easily seen by Peregrines (Gary Thoburn).

Snipe *Lymnocryptes minimus* and Common Snipe *Gallinago gallinago* (Fig. 5.31) in the diet all pointed to one thing – that these birds are migrating at night, and the Peregrines are hunting them during the nocturnal hours. And in countries like Poland where species such as Corncrakes and Spotted Crakes are more common, they appear in the urban Peregrine diet more frequently. Even small birds such as Dunnocks appear commonly in the diet of Peregrines. These secretive, ground-feeding birds are unlikely to be taken by Peregrines in their usual locations such as gardens. However, they migrate at night and it is during this period that they are most likely to be caught and eaten. Other ground-feeding species that migrate at night include Skylark *Alauda arvensis*, Redwing (Fig. 5.32), Brambling and Turtle Dove. The urban Peregrine isn't just a diurnal bird of prey; it is a nocturnal raptor, too (Fig. 5.33). And while it may not have the facial disc of an owl to allow it to hear its prey, the Peregrine can still rely on sight and its big, forward-facing eyes for a night on the town.

It is clear that Peregrines are not venturing out of urban areas to catch these birds, but that these species are flying over towns and cities, mainly at night. Urban Peregrines watch for potential prey in the shadows (Fig. 5.34). The light from street lamps shines up into the night sky and makes the birds moving over visible. The Peregrines then fly up, perhaps only tens of metres, to catch their prey before returning to their perch, often with the bird still alive. In towns and cities with urban gulls, it is easy to see these white birds when they are flying around looking for leftover kebabs and burgers in the middle of the night. Even those flying high can be made out with the naked eye. While the prey of Peregrines may be smaller and darker, their body forms can still be picked out easily against the night sky thanks to the orange or white glow from street lamps. Many of these nocturnal birds may also be drawn to the city lights, mistaking them for exaggerated forms of the moon and stars and becoming disorientated.

In Derby, England, video cameras trained on the city's cathedral recorded the first footage of Peregrines bringing back live Teal, Common Snipe and

Figure 5.33 As the light fades, the Peregrines wait for nocturnal migrants to fly overhead (Sam Hobson).

Figure 5.34 A Peregrine known as Bobbin looks out from her perch in the evening in Worcester (Dave Grubb).

Woodcock in the middle of the night. Meanwhile, in New York, birders watch from the Empire State Building to count the annual migration of flickers, warblers and cuckoos flying over the city at night. Accompanying them are often up to four or five Peregrines catching these small birds as they pass over. This nocturnal hunting behaviour has been directly observed or recorded through prey remains in towns and cities all across Europe, Asia, North America and South America, and appears to be universal rather than restricted to just one population or subspecies. For example, in La Plata, Argentina, urban Peregrines eat a wide range of similar species which lead nocturnal or secretive lives, including White-tufted Grebes *Rollandia rolland*, Spot-flanked Gallinules *Gallinula melanops*, South American Painted-snipes *Nycticryphes semicollaris* and Yellow-breasted Crakes *Poliolimnas flaviventer*. Nocturnal hunting has been recorded on camera at Kaoping Bridge in southern Taiwan, where Peregrines feast on Slaty-breasted Rails *Gallirallus striatus*, Slaty-legged Crakes *Rallina eurizonoides*, Baillon's Crakes *Porzana pusilla*, White-breasted Waterhens *Amaurornis phoenicurus*, Moorhens (Fig. 5.35),

Figure 5.35 The Moorhen is a common nocturnal migrant and is caught by Peregrines all around the world (Gary Thoburn).

Coots and Greater Painted-snipes *Rostratula benghalensis*. Equivalent species to those taken in the UK, such as American Woodcock *Scolopax minor*, Virginia Rail *Rallus limicola* and Wilson's Snipe *Gallinago delicata*, are also eaten as they make parallel journeys during their spring and autumn migrations to and from parts of North America. Diet analysis of Peregrines living in cities in Australia has also revealed grebes, rails and quail species in the mix, and the Peregrines' night-time antics have been recorded on camera.

These nocturnal prey species have many things in common. They are secretive, often crepuscular or nocturnal, migrate at night, and have relatively short, rounded wings and tails. With a quick flapping flight but poor manoeuvrability, they are easy targets for Peregrines. Many are species associated with water and mud, and as an adaptation for these habitats they have evolved countershading plumages, with pale underparts and darker upperparts which keep the birds camouflaged against predators in a watery habitat. However, this plumage combination works less well when flying over cities at night. Their pale chests become illuminated from below, while from above their dark bodies show up against a lighter background. Interestingly, the Coot is an uncommon prey item in the diet of urban Peregrines, perhaps because the species is generally darker-bodied and less easily spotted than Moorhens and Water Rails, which are taken more regularly. In England, thousands of Coots gather in lakes and reservoirs in the west from other parts of the country and abroad, and yet they are absent from the diet of Peregrines in the nearby cities of Bath and Bristol and towns such as Taunton which they may fly over during their journey.

Figure 5.36 Despite the Woodcock being a secretive woodland bird, thousands migrate across our towns and cities at night during the autumn and spring (Gary Thoburn).

Figure 5.37 Water Rails are secretive, reedbed birds. However, at night they fly out of their favoured habitat and migrate out in the open where they are caught by Peregrines (Pete Blanchard).

With the onset of long nights and freezing temperatures, thousands of nocturnal migrants, from Woodcocks (Fig. 5.36) to Water Rails (Fig. 5.37), flee mainland Europe in winter and head west into the UK, with many continuing west to Wales, the Outer Hebrides, Cornwall and Ireland. Suddenly, Peregrines are inundated with prey and at night they can take advantage of the glut, catching, killing and caching prey not just for the next few days, but for the next few weeks. For anyone studying the diet of Peregrines, this can be an exciting time, and daily visits to roost sites can reveal a large diversity of prey species, from Little Grebe to Skylark and Woodcock to Teal. My friend Adele Powell studied Woodcock for her PhD at Oxford University. I sent her some feather samples from Woodcock that had been eaten by urban Peregrines in Bath and Exeter. By using stable isotope analysis, Adele was able to tell me that these Woodcock had originated from Scandinavia and Russia! The ratios of stable isotopes in the feathers match the ratios found in the environment where the Woodcock live, and by matching the stable isotopes to a region, Adele was able to work out where the Woodcock had come from. The results are further supported by satellite-tracked individuals.

These birds migrate at night for a variety of reasons. As well as hiding them from predators such as falcons under natural conditions, this period provides a cooler, humid, less windy environment to fly in. This reduces the amount of energy needed to fly, as well as the risk of overheating and dehydration. It also means birds can optimise their feeding during the day when it is light, and spend night-time on the move. After millions of years of the evolution of nocturnal migration, the advent of electricity and the fitting of street lamps and other lights has happened relatively rapidly in comparison. It now means that birds not only have to avoid predators during the day but also dodge Peregrines at night.

The different types of lighting used in our towns and cities may affect the success rate of Peregrines. There are both narrow-spectrum and broad-spectrum lamps that give off different types of light. The traditional orange or green-yellow lamps provide light only from a narrow spectrum, and may

Figure 5.38 A cached Woodcock left on a gargoyle in Derby (Nick Moyes).

allow Peregrines to spot their prey against the dark sky whilst being almost invisible. However, the new broad-spectrum lamps emit light from a wider spectrum, and may enable prey species to spot the hiding Peregrines more easily.

The ability of Peregrines to take advantage of sudden appearances of prey species also applies away from the main migratory periods, for example throughout the winter when temperatures plummet and snow suddenly appears across the landscape. For ground-feeding birds, their food is suddenly covered over and inaccessible. Therefore, hundreds of thousands of birds, if not millions, from thrushes to larks and Lapwings to gulls, travel away from the snow to find areas free of snow and ice. During such a period, the countryside and urban areas suddenly empty of birds. Many ducks and wading birds may move from southern England to France and Spain overnight. For example, the Golden Plover is a common prey item for the urban Peregrine during winter. In the winter of 2012 one was caught and ringed in a field by ringers based in mid-Wales. A satellite-tracking device was also attached to the bird. During the same week, very cold weather and snow struck this part of the country, and by the weekend the Golden Plover had quickly moved south to Madrid, Spain. Birds normally hidden in woodland, grassland and stubble fields suddenly become very obvious against the grey skies as they depart for areas devoid of snow. Peregrines are quick to take advantage; during such a time they have birds which are easy targets as they are hungry, weak, and very obvious as they fly over in large flocks. Peregrines want to ensure they have enough food to see them through the cold period once all these birds have moved on, so they will be busy chasing, catching and caching many as they pass through.

The autumn migration period and cold spells of weather are an excellent time to check urban Peregrine sites and look for feathers, heads and other signs of prey. Many birds they catch may go unnoticed as Peregrines replenish their cache, storing tens of birds there rather than actually plucking or eating them (Fig. 5.38). The cached food will cool or freeze, and the cache will act as a freezer or refrigerator for the Peregrines. Some food in large caches may go uneaten if there is plenty to eat, and long term may just be left to rot. Either way, for the Peregrine having

Figure 5.39 Staying on the water by day, thousands of Little Grebes migrate at night to new lakes and rivers as they escape freezing weather during the winter (Gary Thoburn).

Figure 5.40 Black-necked Grebes only migrate at night and are occasionally spotted and eaten by Peregrines (Gary Thoburn).

the food there is a safeguard. A squirrel or Jay may do just the same, burying hundreds of nuts even though only a proportion of them may be eaten during the winter months.

Peregrines will also employ some unusual hunting techniques from time to time. In the UK, Little Grebes (Fig. 5.39) are often eaten, alongside the occasional Black-necked Grebe (Fig. 5.40). These are strictly nocturnal fliers, and most are probably taken in cities at night. However, grebes are also excellent divers, and use their underwater environment both to feed and to avoid danger from above. They inevitably have to come back up to the water's surface for air, and it is then that a Peregrine may catch them. Remarkably, Peregrines will fly over the water and hover for short moments over the surface waiting for small grebes to resurface. If they are lucky, they may grasp one with their talons before the grebe knows what has happened.

Attacking other birds, and being attacked

In remote locations on the tundra in the far north of Russia, pairs of Peregrines nest on the ground, often very close to single pairs of Red-breasted Geese *Branta ruficollis* incubating eggs. When an Arctic Fox *Vulpes lagopus* sniffs the ground and heads towards the two pairs of birds, the geese let out an alarm call which alerts the Peregrines. In return, the falcons act as bodyguards for the geese and their eggs or chicks and help to fend off the fox. On other occasions, large Glaucous Gulls *Larus hyperboreus* or Rough-legged Buzzards *Buteo lagopus* may pass by and get the same treatment.

This wonderful, mutually beneficial relationship between the Peregrines and the geese has evolved over time to suit both species, and as such the geese don't often get predated while nesting.

While urban Peregrines may not have a Red-breasted Goose on hand to act

Figure 5.41 An adult Raven in flight.

as a predator alarm, they do react in similar ways to would-be threats or predators. To some degree, urban gulls in European cities give out distinctive alarm calls when a Buzzard or other raptor is looming overhead. However, gulls themselves are also seen as a threat by Peregrines. At the site I have studied in Bath, the Hawk and Owl Trust have frequently recorded injured gulls that have ventured too close for comfort to the Peregrines. Some falcons may go one step further and actually kill wandering raptors that fly into their immediate territory. This has certainly been the case in recent years in Exeter, Devon, where the pair, especially the female, targets mainly juvenile Buzzards flying across the city. Nick Dixon has recorded many instances of them being dive-bombed, and surprisingly the Buzzards have been killed or left for dead on many occasions. This is quite an extreme case study, and usually Peregrines leave Buzzards alone. Robin Prytherch, who has studied Buzzards near Bristol for over 40 years, has only seen a Peregrine strike a Buzzard once. The Buzzard circled round and disappeared into some bushes. The Hawk and Owl Trust's Chris Sperring has also seen this happen once, in 1996. More usually Peregrines simply chase off intruding raptors. In Malaysia, Peregrines won't hesitate to see off Goshawks, Black Eagles *Ictinaetus malayensis* and migrating Black Bazas *Aviceda leuphotes*. Eagles, Buzzards and gulls are all possible predators of Peregrine chicks, and in some cases may attack the adult falcons too.

Ravens have also been grounded in similar attacks by Peregrines, but most attacks don't tend to result in any injury. Judith Smith, founder of the Manchester Raptor Group, researched the interactions between Ravens and Peregrines (Fig. 5.41). Her own observations, alongside those of others, reveal that Ravens do steal the eggs of Peregrines, and no doubt the Peregrines want to ensure they don't lose an entire clutch of eggs to a corvid.

Despite this, Peregrines and Ravens are generally found in close association in rural and some urban locations, such as Bolton, Manchester and Rochdale. While there may be some costs to both species, generally it is better for Peregrines and Ravens to live in close association. Ravens nest on average 1.7 km away from Peregrines, roughly the same distance as the Red-breasted Geese in the Arctic. By living so close, Peregrines can nest somewhere where, as with the geese, they will get early warnings of danger from the Ravens. They can also choose from a variety of unused, ready-made Raven nests – Ravens make many across their territory. Meanwhile, Ravens have a fast predator to help see off any danger, and get to sneak in and steal the odd Peregrine egg or chick from time to time. However, there is some variation in the tolerance of Peregrines to Ravens. Some Peregrines will happily live side by side with them, while others will be the neighbours from hell and persistently attack the Ravens.

CHAPTER SIX

How to Study Peregrines

It is an exciting time to study and enjoy Peregrines. Within discrete populations we are discovering who's who, who's travelling where, and the more intricate details of pairings and territories – in essence, thanks to the help of colour-ringing, satellite tagging and other techniques such as DNA testing, we are discovering that all is not what it may seem. At a fast pace, we are discovering more about the now not-so-secret life of the Peregrine. Each pair in a town or city offers its own soap opera story, followed on a daily if not hourly basis by fans across the country and around the world.

Studying urban Peregrines is a real opportunity to share information, collaborate with others and work together to build a bigger picture of what Peregrines are doing and how best to study, research and protect them.

Studying the diet of urban Peregrines

Working out what an urban Peregrine has been eating is like being a detective. It requires you to examine the evidence carefully, look for tiny clues that perhaps others would miss, and have an open mind to the possibilities of what the dead remains may be from.

Peregrines eat a wide range of bird species, and identifying what is found at or below a feeding site can tell us a lot about their diet. Studying the diet of Peregrines can include monitoring the specific prey items being delivered to a nest or roost in

Figure 6.1 Most Peregrine pellets contain pigeon feathers and are not that useful when studying prey items (Nathalie Mahieu).

Figure 6.2 Occasionally pellets may contain the feathers and parts of other birds, in this case the toes of a medium-sized bird (Nathalie Mahieu).

real time or in recorded time sequences; teasing apart the pellets regurgitated by Peregrines; and finding feathers, wings, skulls, legs and corpses below a Peregrine roost, at a cache, or at the nest. Observing food as it is brought in by Peregrines is time-consuming and may not always lead to the identification of prey. Collecting prey remains from the ground, however, is much easier, although there can be a bias towards larger prey items being found and it is still a snapshot of what urban Peregrines are eating. Generally, I find Peregrine pellets are of limited value (Fig. 6.1). Unlike owl pellets, which contain lots of bones and fine hairs that can be identified, Peregrine pellets contain mainly just feathers, usually from pigeons, that tend to be very powdery or non-identifiable; their structures are badly damaged by the Peregrine's stomach acids. Pellets tend to reveal a much lower number of species compared to simply finding the prey remains. However, it is worth still checking them alongside collecting the parts of birds which the Peregrines drop. The pellets will sometimes contain the leg of a bird, a ring, or may have particularly identifiable feathers of a bird such as a Common Snipe, Water Rail or finch that may not have otherwise been found through other remains (Fig. 6.2).

The regularity of checking for prey remains, local scavengers such as cats, rats and foxes, street or building cleaners and the weather can all determine how many feathers and other body parts you find at an urban Peregrine site. Building up a rapport with the building managers or occupants can help ensure prey remains are put to one side and not swept away. Larger items such as ducks or small geese are generally eaten where they were killed rather than being carried across a town or city.

Regular forays to a roost, either daily or weekly, provide an opportunity to find a more accurate cross section of what Peregrines are eating. It helps ensure that small feathers are collected before they get blown away and avoids a bias towards larger feathers, heads and legs (Fig. 6.3). Daily collections allow for small birds to be better accounted for, though weekly collections will still reveal small feathers of passerines, often clinging to a spider web at the base of a building wall or tugged into the soil by a worm.

Figure 6.3 Prey items that are weeks or months old can be identified and cleared away (Dave Pearce).

Figure 6.4 The heads of birds are nipped off by Peregrines and often drop to the ground. This photo shows those from a Teal (back) and a Woodcock (front).

Figure 6.5 Often the wind catches uneaten or partially eaten birds, which fall to the ground. In this case a Whimbrel has fallen onto the top level of a city car park (Sam Hobson).

Feathers and other parts of prey may be found 50–100 m away – so it is worth searching for items on a wide scale and on the approach to a building. Scrutinise the crevices of paving stones, tufts of grass, drain grills, gutters, soil, curbstones and leaf litter. Safely checking adjacent roads is also useful – legs, heads and bird bodies are often flattened by cars, but still yield information in terms of what species they are from and how many individuals have been eaten. Peregrines will often remove a bird's head from the rest of its body, and the heads are usually found with the cranium opened up (Fig. 6.4). The Peregrines feed on the nutritious fatty acids that form the brain and discard the rest of the head. Sometimes whole corpses will fall to the ground, often blown off a roof or gargoyle by the wind (Fig. 6.5).

Some prey remains may be old – legs and feet may turn to bone, or be in various states of decay. However, they can still be identified even if they are then disposed of. For example, the size and shape of feet, legs and toes can still make them relatively easy to assign to a species, and skulls can be identified with further measurements and comparisons. Skulls from commonly found thrushes (Redwing, Fieldfare, Song Thrush and Blackbird) can be tricky to separate at first, but comparing

How to Study Peregrines   107

the size, shape and colour of the beak sheath (if present) can help distinguish them in most cases. Once a site is being studied for prey items and older items have been removed, anything found thereafter will generally be fresh and new (Fig. 6.6).

I am often sent photos of Peregrine prey remains and these days can quickly work out what bird they are from. Correspondents are often surprised, and are curious as to how I know. Assigning feathers and corpses to a particular species without instant identification features, such as a head, comes with time. When I am helping others to identify feathers, it is repetition and familiarity that enable them to gradually identify the feathers of common birds, and then the more unusual ones too.

Figure 6.6 The remains of a Starling (Dave Pearce).

To identify prey accurately and assign it to a species, the heads, wing and tail feathers provide the most information (Fig. 6.7). However, my friend Luke Sutton researches the diet of Peregrines along the cliffs of Devon and Cornwall and mainly finds the breast and body feathers of birds rather than their main wing and tail feathers. While this makes identification more challenging, with some patience and time they can in most cases still be assigned to a species or a genus. Some birds, such as Common Snipe, Water Rail, Moorhen, grebes, Kingfishers and finches, have distinct body feather textures or colours that can be identified even without the wing or tail feathers.

There are a wide range of ways to identify feathers, and as with anything it comes with practise. Resources are more widely available now as identification of feathers and other parts of birds becomes even more important for understanding

Figure 6.7 Various wings and corpses laid out that have been collected from an urban Peregrine nest site.

the lives of both common and endangered wildlife. Books, the Internet and museum collections all provide references which together can help build a comprehensive examination of what the Peregrines have been hunting and eating. For birds in the UK and across Europe, my favourite book to use is *Tracks and Signs of the Birds of Britain and Europe* by Roy Brown, John Ferguson, Michael Lawrence and David Lees. The beautiful artwork of feathers and skulls provides a valuable tool to begin identifying prey remains. Other useful books include the photographic guide *Feathers: Identification for Bird Conservation* by Marian Cieślak and Bolesław Dul, which focuses on the feathers of raptors but also features species with similarly patterned feathers, such as wading birds. *Moult and Ageing of European Passerines by* Lukas Jenni and Raffael Winkler is also a useful reference book for looking up tiny details on a bird's wings, and for looking at species with very plain, nondescript feathers which are not covered in the other books. Online, a search for the *Atlas of Birds of Aragon* provides a chance to see a range of birds in the hand, including their wings and body plumage. Meanwhile, the Dutch ornithologist Michel Klemann has a brilliant online feather database of most birds you would expect to find in Peregrine prey – just search for his name and the word 'feathers'.

What about further afield? In the USA, the book *Bird Feathers: A Guide to North American Species* by S. David Scott and Casey McFarland is a useful feather identification guide for bird remains in North America, while the online *Feather Atlas: Flight Feathers of North American Birds* produced by the U.S. Fish and Wildlife Service provides photos of a range of birds.

In Japan, the book *The Feathers of Japanese Birds in Full Scale* by Masaru Takada and Takuya Kanouchi provides the most comprehensive guide to bird feathers, with high-definition photos and colour pages showing a range of feathers from species found in Asia, as well as many which are found in Europe and extend their range across to Japan.

Meanwhile, a *World Feather Atlas* series is currently being produced by the German Feather Research Group and will be available in the future – more information on these resources is at featherguide.org.

How do you know if the feather has been plucked by a Peregrine or simply been moulted?

The tips of feather quills normally give a clue as to whether the bird in question has been plucked by a Peregrine or has moulted the feather. Unmoulted feathers are still connected to the wing tissue when in the bird's wing. If they are plucked, the tissue and adjacent blood vessels are torn, resulting in a dark tip to the quill where the blood has dried. Additionally, the feather webbing may be torn where the Peregrine has gripped and pulled, and the quill itself may be indented or bent. The quill tip that would have formed the attachment in the wing may also taper more narrowly. In moulted feathers, the quills become disconnected from the wing tissue and fall out without any tearing. Plucked feathers may have red blood splattered on them, and are often found in clumps. Many feathers of the same bird or same species may be found across a large area below a Peregrine site. They are likely to be plucked, unless a pigeon, wagtail or other bird roost is close by. Heads, legs, wings and corpses may also accompany the feathers.

How do you know if the feathers are not from a Sparrowhawk kill?

Peregrines and Sparrowhawks may often share the same urban habitats, and therefore their prey items could be confused. However, both generally use different places to pluck their prey. Sparrowhawks will pluck their prey in the place where they brought it to the ground or will take it to a favourite perch in a tree, usually under cover. A pile of dove or Starling feathers is often a good clue that a Sparrowhawk has been feeding in a garden or park. And whilst they may briefly perch on a church or building used by a Peregrine, it is unlikely they will use it regularly for plucking. Meanwhile, an open location, a mass of feathers and body parts, and regular white faecal splats are all very diagnostic that a Peregrine is in town (Fig. 6.8). Peregrines will feed on birds such as ducks *in situ*, but most birds that they bring back to an urban site are plucked and eaten up high.

Figure 6.8 The white faecal splats known as mutes are a telltale sign Peregrines are in residence at an urban location.

Occasionally, urban sites are used simultaneously by Kestrels, which feed mostly on small mammals but will also eat birds such as Starlings. However, the two living side by side in the middle of a town or city is unusual, and Kestrels themselves can become prey to Peregrines. Where I have witnessed the two species living close together on an aeroplane hangar, the Kestrel pellets, faeces and plucked feathers were very focused in one place directly below where the Kestrels were nesting. Those most likely to be plucked by the Peregrines, especially pigeons, were spread all around the building, and reflected the Peregrine's nature of using lots of different perches and plucking its prey up high. Kestrel pellets are smaller than those of Peregrines, and mostly contain fur and bones rather than feathers.

A step-by-step guide to collecting and sorting prey remains

When studying urban Peregrines there is often the opportunity to look more closely at what they are eating. This can be useful for writing scientific papers for journals, papers for regional bird reports, and for telling the public more about the Peregrines. The guide below outlines the key steps to follow to gain the most accurate data with regard to prey remains found at or below a Peregrine site.

It is worth noting that in many countries it is illegal to be in the possession of bird feathers due to the protection birds have, and permission may need to be sought from the appropriate authorities.

Figure 6.9 Female (top) and male (bottom) teal body feathers lying on the ground below a Peregrine roost (Hamish R. Smith).

Step 1

When looking for prey remains, scrutinise the ground for feathers, both around the building and up to 100 m away. Feathers can fall or be blown some way. Gloves are advisable as prey can be rotten and wet, ideal for bacteria and germs. Look for small feathers as well as larger wing and tail feathers (Fig. 6.9). Collecting all of the feathers allows the number of individuals involved to be determined; for example, if you find two of the same outer wing feathers from the left wing of a Feral Pigeon, you know two birds have been eaten.

Step 2

Feathers, heads, wings, legs and anything else can be collected in a bag and taken away (Fig. 6.10). Removing feathers and body parts from the site means any new fall of feathers and other prey remains doesn't get confused with those that have already been recorded. Feathers that are damp or wet will dry on absorbent kitchen towel or newspaper. Any fleshy parts you wish to keep, such as heads, can be put into a sealable plastic bag, labelled and frozen. Legs can be air-dried, although it is worth cutting off any fleshy bits or tendons at the severed area. If collected during the spring or summer, bird heads can be left outside in a container for flies to lay their eggs in and maggots to clean the skull. You can also bury bird heads, although this tends to make the internal spaces of the skull dirty as microorganisms take mud into the spaces in the bone that are hard to clean. In a laboratory, a less labour-intensive alternative to boiling skulls is to use sodium perborate tetrahydrate. The chemical works well at removing protein, dissolving any flesh and skin. However, any bill sheaths left on the skull will also dissolve. This process leaves the skulls looking very clean and unnaturally bleached a yellow-white colour. Some laboratories may also use colonies of *Dermestes* beetles, which are very effective at cleaning skulls but also create a very smelly room!

Figure 6.10 After collection and bagging (top), feathers can be laid out and separated according to species and individuals (bottom).

Step 3

Feathers can be separated out into their different species and further separated into which part of the body they are from (Fig. 6.10). For example, with pigeons it is important to look carefully at the primary and tail feathers for duplicates of the same feathers. Feral Pigeons come in a range of different colours. The outer tail feathers of blue-grey Feral Pigeons usually have a white outer web, while the outer wing feathers are distinctly curved with

Figure 6.11 Some feathers may need closer scrutiny to identify to a species.

a thin, stiff outer web. This can help determine the number of individual birds you have. Primary wing feathers from different pigeons may also vary in their shade of grey. Additionally, Feral Pigeons may be white, black, dark blue-grey, red-orange, or white and grey, as well as various mismatches of colours! The number of wings, rings (if racing pigeons) and humeri (wing bones) may also help determine the number of individual pigeons eaten. Many body feathers of birds can be fairly nondescript between species and make identification difficult. However, over time subtleties between close species can be made out. For example, the primary feathers of Song Thrushes have a brighter yellow-orange inner web compared to those of Redwings, which are generally a dull, off-cream colour, and lack the deeper hues.

Some feathers and skulls may need further comparison with the real thing, and a visit to a local museum may be in order (Fig. 6.11 and 6.12). Museums that deal with biological specimens will have drawers containing study skins. These are birds that have been preserved and laid out straight with a stick coming out from their rear end so they can be picked up without a need to touch them directly. Mounted specimens, those which look very real and are in their natural poses, may also be useful, and some museums have handling specimens which can be held and examined in closer detail.

Figure 6.12 Unusual wings and feathers may sometimes be found, such as this Great Grey Shrike.

Step 4

Once the prey remains have been assigned to species and the number of individuals counted it is important to keep a log of this information, with dates and the location where they were found, in both a notebook and an electronic format such as a spreadsheet. Both provide a backup for each other, and the electronic data can be easily formatted and analysed. For the electronic format the data can be presented as daily, weekly and annual totals of each species. These help to present certain trends and compare between certain time periods. It is good practice to have as much detail as possible, even if the data may not seem relevant to you. Some Peregrine diet studies often just assign prey to the families of birds they are from, but if they can be

Figure 6.13 A range of Peregrine feathers which are frequently found below an urban Peregrine's roost or nest site. From left to right back: juvenile male tail, adult female tail, adult male tail, adult female secondary, adult male secondary, juvenile alula, adult male inner primary, and adult male outer primary. From left to right front: underwing covert, underwing covert, tertial.

identified to species level then this is more accurate and provides a more interesting insight into the lives of Peregrines and their prey, not only for the study but also for local bird recorders, county bird reports and other researchers. If someone else wanted to use the data in the future, it also means they have a wealth of information, rather than just a restricted data set.

Peregrine feathers

A Peregrine's feathers are frequently found below a roost or breeding site (Fig. 6.13). They are a good way of discovering if Peregrines are present at a site, especially if prey remains are not evident or are scarce.

The most common feathers are those from the breast and chest – they are white with dark grey lines. The lines bleed off into the white, as if someone has drawn them with graphite or charcoal and smudged them with their fingers. The white areas may also be creamy and dirty looking.

The long primary feathers are slate-grey with inner webs of alternating dark grey and smoky, off-white bars. Unless you find them fresh, moulted

Peregrine feathers are often weather-beaten, dirty and bedraggled, but can still be identified as belonging to a Peregrine. The tail feathers are grey all over, with darker bars and cream or white tips. In juvenile birds the dark grey colours are replaced with dark brown pigments in the primary and tail feathers.

The clear size difference between male and female Peregrines means any moulted feathers can give you a clue as to which genders are present at an urban site. If both are present, then the male tail and wing feathers will be clearly smaller than the female's. For example, in the nominate race of the Peregrine *Falco peregrinus peregrinus*, the male tail feathers will be on average 144 mm long and the female's 173 mm long. With a little time and experience the differences between male and female feathers become more obvious just by looking at them.

It is worth keeping the feathers of Peregrines, both out of interest and also in case DNA tests or studies are required to look at genetic diversity or to confirm the origin of individuals either in the wild or captivity. The feathers are ideal candidates for this. They can be stored in dark containers so the light doesn't damage the DNA. The opportunity is also arising for the feathers of prey items found at nests or below sites to be identified by DNA analysis, if identification is otherwise unattainable.

Providing a nest box

Studying urban Peregrines provides opportunities to observe them up close in ways that haven't been possible with falcons nesting in remote, rural locations. Nest boxes are an ideal way of observing Peregrines in a convenient location, but more importantly provide a safe place for Peregrines to lay their eggs. If you are going to make and put up a nest box for Peregrines it needs to be for a good reason. Perhaps a box is required to replace part of a building or roof which has become inaccessible or pulled down; it may be that a pair has laid onto a roof and the eggs have just rolled into the gutter; or perhaps a pair has been resident on a building for several years but doesn't have an obvious platform, nook or cranny to

Figure 6.14 This nest box design is perfect for an enclosed flat roof, providing cover from the rain and sunshine, and easy access for the chicks to explore.

Figure 6.15 This nest box is more exposed to the sunshine, and therefore has its sides and back facing south-west to protect the chicks from the heat of the day.

lay eggs. They may also just be biding their time. However, this does not mean Peregrine nest boxes need to be put anywhere and everywhere, as they are unlikely to all become occupied and the effort should not to be solely tokenistic.

Nick Dixon and Colin Shawyer have produced a wonderful leaflet that outlines everything you need to know and what to keep in mind when considering an artificial box, and whether putting one in place is really necessary. For some buildings a simple modification to a ledge or cavity may be all that is required. A quick search online for Peregrines and nest sites will bring up the leaflet, which is called *Peregrine Falcons: Provision of Artificial Nest Sites on Built Structures*, under the London Biodiversity Partnership's website. It is worth asking whether a nest box is really needed, and why it would be worthwhile putting one in place. Are the Peregrines finding it difficult to successfully hatch eggs? Is the Peregrines' usual nest site at risk of being demolished? Have the Peregrines been in residence for a few years or more and shown courtship but no egg laying? Nest boxes are generally site specific and there is no general design which fits all.

However, one Peregrine nest box design I favour for a flat roof is one used on a church in Cheltenham (Fig. 6.14) and on an office block in Bristol (Fig. 6.15) – it is a good size for all the Peregrine family, sheltering the adults and young during hot sunshine and providing cover when it is wet. Put simply, though, a large raised box tray with gravel will suffice. It is all about replicating what a Peregrine

Figure 6.16 In Germany, baskets are installed on industrial buildings (top) and pylons (bottom) for Peregrines (Thorsten Thomas).

Figure 6.17 This nest box in Derby overhangs the side of the Cathedral and provides lots of space for the chicks to move around and flex their wings prior to fledging (Nick Moyes).

would use in the wild. Peregrines tend to nest in a variety of places, but often parts of cliffs under overhangs are favoured. Cover isn't essential, but in urban locations where freak downpours can suddenly flood a roof, a raised box with a roof helps protect the eggs and chicks in the absence of rocks and tree branches. In Germany, dog baskets with gravel are fixed to the flat walking areas of industrial buildings (Fig. 6.16).

An artificial nest box being planned for attachment to the side of a building with a drop below will require a platform or extended box space for the chicks to move around (Fig. 6.17). It will need to cater for four or five chicks flapping, squabbling and moving around close to fledging. Unlike with a box on a flat roof, the chicks won't be able to just walk out and explore an extended flat surface. With an additional roof, gravel substrate, a large area for movement, and a decent sized lip to the platform sides the chicks have a good chance of fledging successfully and safely.

In terms of sizes and dimensions, because nest boxes are site specific it is better to get additional advice from raptor groups or experts close to you or online about your specific needs and location.

The Internet and web cameras

In Sussex in 1998, the first Peregrine nest to be broadcast around the world from the UK was sent out across the Internet. Under the close eye of researcher Graham Roberts, the image of small white chicks in a nest box was grainy and updated every few seconds or minutes, but nonetheless people were able to see a live Peregrine nest from their office desk or living room. Only one year

Figure 6.18 In Bath the BBC installed this high-definition camera during the broadcast of the wildlife series *Springwatch*. A tiny camera inside the nest box helps monitor the eggs and chicks long term (Hamish R. Smith).

Urban Peregrines

Figure 6.19 In Aylesbury, Buckinghamshire, the nest box on the county hall has a camera installed on the roof of the structure looking down on to the nest (Mike Wallen).

earlier, the first ever live footage of nesting urban Peregrines had been broadcast across the web from a nest on the Sun Life Financial Centre building in Toronto, Canada. Both were the very beginning of many people's enjoyment, fascination and hobby (even bordering on obsession!) of watching Peregrines online from nests around the globe. As an example, in Derby in the East Midlands, where a pair of Peregrines nest in the city centre, the web cameras have been viewed 2.5 million times since they were installed in 2007, and in the first half of 2013 alone there were 300,000 hits as people followed the lives of the Peregrines from all across the world.

In the early days, watching Peregrines online required some patience as live pictures were updated every few minutes and web pages would suddenly freeze! Fortunately, with the roll-out of broadband (rather than dial-up), and smaller, cheaper and more reliable technology, cameras are now able to stream high-quality, real-time pictures of Peregrines online. Not all cameras set up at Peregrine nest sites are streamed across the Internet. Some may just be connected to a monitor to show local people, for example in churches or offices, and researchers may use small screens or monitors to remotely study a nest without a need for lots of cables, computers and additional costs (Fig. 6.18 to 6.21). Either way, we are now able to watch Peregrines in immense detail from the time they display to when they lay their eggs, hatch their chicks and fledge their young (Fig. 6.22). It also allows those studying Peregrines to remotely check on the birds when nearby building work is

Figure 6.20 In Norwich, Norfolk, two cameras look down on the Peregrine family below (Andy Thompson).

Figure 6.21 In Cheltenham, a security camera provides live footage down to a monitor in the church below (Dave Pearce).

Figure 6.22 A whole variety of behaviours that are missed by usual means of observation, such as regurgitating a pellet, may be spotted on web cameras (Dave Pearce).

taking place or other disturbances occur, allowing swift action to rectify any problems.

The technology for web cameras is changing all the time and in recent years tracking and filming wildlife has become more accessible to everyone, not just those in the television business. Today, very small cameras produce good quality pictures, a far cry from the huge security cameras that were originally used. There are relatively simple and cheap devices on the market, and advice can also be sought from those who already use cameras in Peregrine boxes. Some now have motion detectors and will only film the birds when they move or fly in. However, there are still costs that need to be considered, so before looking at the kit itself it is worth asking why a web camera is needed, what you hope its use will achieve, how it would happen, and whether installing one is actually feasible. Will the pictures be streamed online, or shown locally on a screen? Is there already another site nearby providing the same service? What are the aims and objectives of installing a web camera? What costs are involved? Where will the money come from to do this?

If your objectives are to promote the Peregrines and provide a better understanding of how they live and why they are important, then gaining an insight into their usually private lives using web cameras is a great way of enabling others to enjoy and love the birds. This in turn can translate into behaviour and attitude change, leading to a greater interest in and respect for Peregrines and other wildlife. Watching Peregrines in this way connects people with their 'biophilic' side, the innate nature-loving need and desire to be at one with wildlife which keeps everyone happy, healthy and alive. A document published in 2010 named *Branding Biodiversity: The New Nature Message* by Futerra Sustainability Communications looks at how best to engage different audiences with nature. If you enable people to love nature, once they are inspired they will want to know what to do to help it.

- What resources have you got to show visitors what a Peregrine looks like?
- Have you got equipment which is appropriate for children and adults? (Think about height of tripods and telescopes.)
- How many telescopes do you need? Three or four may be needed for particularly busy sites.
- Will fundraising be part of this event?

Many people may walk past a Peregrine site and have no idea that Peregrines live there. When they are shown the birds through a scope it can be a very exciting experience for them. It is worth remembering that not everyone will know what a Peregrine is, and families will need telescopes set at a low level for children. Telescopes are a great way of showing people Peregrines – though to the untrained eye even a bird in the middle of the view can be difficult to see, especially if it is the same colour as the background. Binoculars are less useful for the public, especially if they are not used to using them – much time can be spent trying to search for something without success, while simply using the naked eye to pinpoint a bird may be more effective. Inevitably some people will give the impression they can see a Peregrine when they obviously can't. It is worth persevering. Once someone does sees a Peregrine you will soon know by his or her positive response!

When showing people Peregrines, or any spectacle, it is worth just pulling back and allowing your audience to enjoy the moment and take in what they are seeing. After they have been 'wowed', there is then an opportunity to put across facts and allow them to reflect on what they have seen. Quite often we talk while the experience is happening, and research shows that the messages we are trying to communicate often get ignored. Pausing while the action is happening and chatting a moment later can make a big difference to the information people take away with them.

Figure 6.24 A Peregrine watch at Norwich Cathedral attracts up to 30,000 people (Dave Gittens, Hawk and Owl Trust).

Liaising with landowners and stakeholders involved with Peregrines

For those with Peregrines on their land or buildings, their presence can be an exciting opportunity and provide some good publicity. Peregrines are generally well received in urban areas, but some PR work may also be in order to relieve concerns by some about mess, blocked gutters from fallen prey remains, and noise.

Whether it be the rail authorities, the priest of a church, a building manager or a consortium of flat owners, liaising with and seeking permission from different groups of people or individuals involved with a building or structure that Peregrines are using will be imperative, both for the Peregrines' well-being and also the opportunity to observe and study the birds. Often it is just a case of seeking landowner permission to enter certain locations. In other situations it may require further discussion at a committee meeting or the submission of a risk assessment prior to entering a space. It may also involve being accompanied by staff during the checking of a nest or ringing of Peregrine chicks, especially if pylons, industrial-type buildings and offices are involved. Either way, the process can be made easier by developing good relations, working with those involved, and ensuring that updates, photos and developments are communicated back quickly after your visits.

Top tips
- Be clear about what it is you want to do. Are you just letting the building owners or residents know that the Peregrines are present? Do you want to study the birds and, under licence, visit the nest? Do you want to collect prey remains? Do you want to install a nest platform?
- Write a risk assessment and method statement and ensure any risks are minimised, so that yourselves, those with you, and the birds remain safe.
- Think about how your visits and studies will benefit local people and the sites which the Peregrines are using, and how you can communicate this. This could range from an article in a newsletter to a poster in a church.
- Consider timings that will suit the landowners or stakeholders rather than just your own timetable – this may mean very early starts or evening work. If climbers are involved, weekday evenings may suit better than weekends when they are away climbing.
- If ringing chicks, or photographing the Peregrines, take photos which can be sent back and distributed to staff of sites, or to tenants or flat owners.

- Where possible, involve those who have given site permission or need to be on location with you in helping with carrying kit, or in writing down details such as biometrics when you are ringing Peregrine chicks.

Risk assessments

Risk assessments are an important part of assessing a situation prior to ringing Peregrines, setting up a nest box or collecting prey. They don't have to be onerous or long, and mean that the activities, which may include others or be on behalf of an organisation(s), are safe – we all have a responsibility for those around us. In the UK, the Health and Safety Executive, hse.gov.uk, has detailed information on risk management. The key things to consider are the potential hazards and risks, those at risk, a risk rating (how likely is it to happen) and mitigation (how the hazard or risk can be minimised or removed). A risk assessment will differ according to the site, activity and people involved.

For nest-related work such as ringing Peregrine chicks and installing web cameras, the safety of both the birds and the climbers is essential. It is important to consider the experience of the people you are using to do this, and to ensure that all health and safety has been taken into account. I am not a climber or an abseiler myself, and therefore work closely with volunteers from the British Mountaineering Council for advice and to reach Peregrine nest sites under my supervision and licence. A team of highly experienced climbers and abseilers works effectively together, has the correct equipment (and backup), and knows exactly what to do if anything is to go wrong. They also produce their own risk assessments for the operations, and help advise me on any which I may need to submit to landowners or site managers.

Chapter Seven

Ringing Urban Peregrines

There is something incredibly special about seeing and holding a young Peregrine. For me, their special status as a Schedule 1 bird species means that having the privilege to find out more about them up close is even more important – they are not a bird just anybody can handle. I began ringing Peregrines, the process of putting a metal and a colour identification ring on their legs, in 2007 when my friend Ade George trained me up. We ringed two chicks in Bath that year. I remember feeling so nervous and shaky. Recently, while training up others, I observed their own hands shaking and could relate to how they were feeling. But equally I could see how much they were enjoying seeing a bird they were so passionate about up close and personal. We don't handle Peregrines or visit urban Peregrine nests casually – they require an experienced nest recorder or ringer with a Schedule 1 Licence to be present, and the procedure of visiting a nest often requires a lot of coordination with landowners and other stakeholders.

Across the globe we have an incredible amount of knowledge and data about the movements of birds thanks to bird ringing. The technique of ringing (or banding) a bird has been used for over 100 years. After a bird has been ringed, it may at a later stage get recaptured or be found dead. The information from such a recovery opens up a whole new world into our understanding of how long birds live for, where they move to and from, and how their populations are faring over time. By following particular individuals we can also find out how many chicks they have produced during their lifetime, about the partners they have been with, and the nest sites they have chosen to defend and breed at. Over recent years this has also provided more information relating to extra-pair copulations in Peregrines, as well as divorces, inbreeding and cooperative breeding. Not everyone may be on board with ringing to begin with. There may be uncertainties regarding what it means for the chicks (and the parents) in terms of their welfare and how rings interfere with the natural image of a Peregrine, especially if it is being photographed. There may also be a lack of understanding about what ringing is and how it works.

Despite the advent of new technologies, bird ringing still remains a relatively cheap and effective method for tracking birds. To increase the chances of finding a small bird at a later date in a new location, hundreds of small birds need to be ringed. However, Peregrines are large birds, loyal to particular sites, and popular to watch. Therefore, for every ten Peregrines colour-ringed, at least one is likely to be resighted, be it dead or alive. If they were just given metal identification rings this figure may be much lower – perhaps 1 in every 50 or 100 birds may recovered, if that. Metal rings are very difficult to read and often it is only once the Peregrine is dead that such a ring can be read or recovered. Peregrines by their nature are relatively easy to spot and observe. And with the advent of digital photography, many resightings of colour-ringed birds are recorded through people taking chance photographs rather than spotting the birds through a telescope or finding them dead, although both still make important contributions to the data.

Developments in technology now mean that alongside ringing, a variety of birds can also be identified and followed with colour-coded wing tags, neck collars, colour leg rings, satellite transmitters, radio-tracking transmitters, geolocators and Passive Integrated Transponder (PIT) tags. The method used often depends on cost and on the size of the bird, to ensure that the identification marker doesn't impede their welfare or survival. The size of Peregrines means that they make good candidates for some of these tracking methods. In the past, radio transmitters have been put on to Peregrines to track local movements, and falconers still use them today to find their birds if they go astray. The advent of cheaper and smaller devices now means that satellite transmitters can be placed on to Peregrines – these provide data which doesn't rely on the researcher physically following the birds, and means wherever they are in the world we still know their location, as long as there is enough light each day to recharge the transmitter's batteries (Fig. 7.1). Peregrines usually sit out in the open so the charging of batteries isn't a problem, but it can be for other species such as Common Cuckoos *Cuculus canorus* and Woodcock, which may hide during the day in dark woodlands or forests. While the technology and the download of data can run into thousands of pounds per bird, it does provide us with accurate data on all the individuals that have been tagged. It is less of a lottery compared to ringing,

Figure 7.1 A juvenile Peregrine in Poland is fitted with a satellite-tracking device (Slawomir Sielicki).

where hundreds of ringed birds may never be seen again! In recent years, the migration movements of birds such as Common Cuckoos, Turtle Doves, Eurasian Nightjars and Nightingales *Luscinia megarhynchos* have all been tracked thanks to satellite technology or similar geolocators which record the daylight hours in relation to the latitude and longitude of where a bird has been living. While the latter devices don't give live readings and rely on the birds being recaptured, the information gleaned about many of the species Peregrines may prey upon is priceless. Peregrines have also been satellite tracked, with studies revealing their migration from North to South America, Siberia to Southern Asia, and local movements of urban birds tagged in European cities such as Warsaw, Poland.

For your average researcher studying Peregrines, colour-ringing chicks and perhaps adult birds offers the easiest and cheapest method of monitoring them. Trained ringers with a Schedule 1 Licence for specific nest locations in the UK are able to access nests when chicks are the correct age and fit both a metal identification ring and a colour ring. Ringing Peregrines in other countries around the world also requires licences, training and official permissions. In some countries, such as Denmark, ringing Peregrines has only recently been allowed. There has also been an attempt across mainland Europe by the European Peregrine Falcon Working Group to coordinate a system of colour

Figure 7.2 Colour rings used to mark individual Peregrines (Sam Hobson).

rings to ensure that there isn't overlap or confusion between the same colour schemes and letters being used by adjacent countries.

In the UK, a metal identification ring issued by the BTO must always be placed on one leg, and the colour ring on the other. The colour rings have letters or numbers, meaning they can be read from a distance through a telescope or when the bird is photographed in flight; zooming in to the image can allow the ring to be read. If setting up a colour-ringing project it is worth bearing in mind the colour combination for these rings – for example, blue rings with white letters will read better than those with black letters. And certain letters should be excluded as they can be easily confused with others – these include E, I, M, O, Q, U and W. The colour rings are issued and coordinated by the BTO in the UK, and specific colour-ringing projects can be registered (Fig. 7.2). Alternatively, a generic pool of colour rings may be available for *ad hoc* ringing of nesting pairs that are not part of a wider study.

As mentioned earlier in this chapter, British Peregrines are classed as a Schedule 1 species. This means that to disturb a Peregrine for nest recording or ringing purposes, a Schedule 1 Licence is required by an experienced bird ringer. This is issued by the BTO and any ringers must refer to the *Ringers' Manual* for clarification on details. The following information is provided as a guide for those in a position to legally ring young Peregrines, and includes my own observations. The book *Raptors: A Field Guide for Surveys and Monitoring* by Jon Hardey, Humphrey Crick, Chris Wernham, Helen Riley, Brian Etheridge and Des Thompson also outlines important details for monitoring and ringing at Peregrine nest sites.

Tips for ringing Peregrines

If the adults are nearby when you approach the nest, they may fly off and perch close by to watch the activity, or they may fly around the building or nest site calling loudly. In some parts of the world, such as the USA, Peregrines may perform attack dives towards the ringers at the nest, although this doesn't seem to be the case in the UK. As the main ringer and licence holder for the nests I ring at, I need to keep my eye on the bigger

Figure 7.3 A three-week-old Peregrine chick is ringed by a trained and licensed ringer.

Figure 7.4 Alongside the ringing, biometrics such as the length of the head are measured.

Figure 7.5 Alongside a metal identification ring, an additional colour ring is carefully placed on the other leg.

Figure 7.6 The Peregrine chick is ringed and ready to be placed back in its nest.

picture, so I am always aware of where the parents are and what they are up to. If they do decide to watch from afar, they often perch on a pylon, tree or church gargoyle.

Ideally Peregrine chicks need to be ringed, measured and weighed in 15 to 20 minutes (if there are around 3 to 4 chicks) to avoid chilling. If the chicks are removed from the nest box itself and taken a short distance away for ringing, it is worth having someone at the nest box itself or covering the nest scrape with some cloth or a bird bag, so that any returning adults don't see an empty nest.

Peregrine chicks should generally be ringed at around 21 to 26 days old – by this point their tarsi are at a thickness where age can be determined, the legs are not going to get much wider, and the rings are not going to slip over the toes of the birds or get stuck on the joints (Fig. 7.3). Any older, and there is a risk that the chicks may move away from the nest box and fall (if there is a drop outside), or scatter if on a flat roof. As with any ringing, we want to create minimal impact and disturbance. The chicks grow fast, and at 21 days it is remarkable just how large they are after hatching from an egg just a little smaller than that of a chicken. By this time, the chicks will have developed their next coat of down and will be in varying stages of wing feather growth. At this stage they can they can keep themselves warm and need little shelter from their parents. On a cold day, however, they can still get chilly and need to be ringed and measured as quickly as possible. Those around 21 days old will just be showing signs of the primary feathers breaking through their pins, the blood-rich tubes which grow from the wings. The chicks at this point in time are unaware of their killing talent, and their talons tend to be soft and relaxed. However, older birds may lie on their back and present you with their talons, although not to the same extent that I have noticed in Kestrels. They may also still have their egg tooth, a tiny, hard white process which sits on the top of the upper mandible. It will usually be fairly obvious whether you have a male or female chick, especially if you have a mix of the two. The females, even at three weeks old or less, are larger than their male counterparts and have thicker, longer legs. The BTO's *Identification Guide to European Non-passerines* gives a set of useful biometrics. Measuring the length of the tarsus will provide a good guide to what gender you have, along with bill length and general look of the bird: a male's leg will be on average 46.9 mm long (minimum 45 mm and maximum 49 mm), while for a female it will be 53.5 mm (minimum 51 mm and maximum 56 mm). When ringing the young birds it is worth getting further measurements such as weight, head length, bill length, bill-to-cere length, leg width, tarsus length, and tarsus-and-toe length (Fig. 7.4). These all provide data which can be used for further studies of Peregrines and also help with deciding whether you have a female or male chick. It is also worth checking whether there are any current studies or organisations requesting moulted feathers for DNA analysis and studies of individuals as well as for providing a bank of data to help with Peregrine-related crimes such as the theft of chicks. Occasionally the feathers direct from birds may be required, but this requires strict permission to be given to the ringer and their methodology for doing it to be clarified.

The chicks themselves will have different temperaments – some may be noisy and fidgety while others may be relaxed and look around with little calling at all. When handling the chicks, watch out for panting or shivering which may mean they are cold, hot or stressed and need returning to the nest as soon as possible. When the chicks are put back into the nest they can be covered over with cotton bags until you or the climbers are ready to move away. The bags can then be quickly removed, leaving the young birds to sit quietly and relax. Schedule 1 Licences in the UK require nest recording information to be submitted, so do record any unhatched eggs. The licence may also allow these to be removed safely in an egg box and sent off to the Centre for Ecology and Hydrology for further analysis.

In Scotland, the Lothian and Borders Raptor Study Group, in association with other raptor groups in Scotland and northern England, has had success using Passive Integrated Transponders to look at Peregrine survival rates and annual turnover of individuals. The transponders are attached to a Peregrine via a colour or metal ring, and are detected by a reader and antenna located at the Peregrine's nest. Unlike with other ringing techniques, the adult birds are captured on the nest. The eggs are replaced temporarily under licence with dummy eggs and kept warm, and a special trap is set to catch the returning adult, who is subsequently ringed and tagged. The eggs are then put back and life returns to normal. With the reader and antenna also in place, any PIT-tagged Peregrines visiting the nest site will be detected and traced.

Key things to remember when ringing Peregrines
- Ringing kit (pliers, rings, calipers, large wing rule, glue, colour rings, pen, ringing book/data log, weighing scales, circlips).
- Schedule 1 Licence and ringing permit.
- Cotton bags to put the chicks into.

- Towel to give the chicks something to grip on to and stay calm.
- Watch/mobile phone for keeping track of how long the chicks have been handled for.
- Risk assessment/method statement.
- Landowner permission and details.
- Informing police of the event.
- Developing good relations with climbers and others involved.
- Taking photos of each chick – useful when replying back to people who have resighted your ringed birds, and for future reference (Fig. 7.5 and 7.6).
- Keeping in regular contact with your team, and clarifying roles both prior to and during the ringing days.

Involving the media

Filming Peregrines and the ringing of their chicks for factual television and news is very popular, and a great way of sending positive messages out about these birds. One of my best mates Mike Dilger, also known on UK television for presenting *The One Show*, *Inside Out* and other programmes, has been a great supporter of my Peregrine research over the years. On a number of occasions Mike has presented pieces for regional BBC television relating to my Peregrine diet work and the Peregrines that I have studied in Bath. Peregrines make good television, and viewers enjoy them. Local champions in television are able to help make stories work for large audiences and help you reach more people (Fig. 7.7).

Springwatch, a BBC-produced series broadcast in the UK, looks into the lives of all sorts of wildlife, from bumblebees to Blue Tits and snakes to shrews. Nest cameras reveal behaviour that has often been previously unobserved, even by those who study the animals in more detail. The urban Peregrines I study were featured on *Springwatch* in 2010 and 2012 (Figs. 7.8 and 7.9). Their accessibility and resilience makes them ideal birds for a web camera set up for a live television show. Like other birds featured by *Springwatch*, many of which may rarely be seen on film,

Figure 7.7 Filming for a short story about urban Peregrines featured on the BBC with Nigel Middleton (Hawk and Owl Trust) and naturalist and presenter Mike Dilger (right).

Figure 7.8 The BBC filming the ringing of the Peregrine chicks in the Avon Gorge in Bristol. Here climbers from the British Mountaineering Council are collecting the chicks. The activity went ahead without any delay to the ringing procedure or return of the chicks back to their nest.

they draw a sense of fascination, awe and wonder. Enjoying something local, just down the road can make a difference to people's lives – and in turn people show a tendency to care more or want to do something that feeds their fascination or appreciation. They feel involved and part of the story.

However, with filming comes the responsibility to ensure that the birds' welfare is put first, and also that suitable messages are being conveyed both verbally and also visually. There are many things that could be misinterpreted by members of the public watching activities on television or online. For example, in 2013 some footage was released from one site showing Peregrines attacking the ringers at the nest. The adults were swooping down and the ringers were using wooden shields to protect themselves. Such film out of context is not useful in promoting bird ringing, and can give the impression that the birds are being harmed and are in distress. In reality, this particular pair are bolder than most other Peregrines and naturally defending their territory. After ringing, the parents would have calmed down and behaved as if nothing had ever happened. But the moment of attack shown as a clip on the Internet gives a very different impression.

Prior to and during filming it is essential to liaise with the television producers, presenters and cameramen so everyone understands clearly what can and can't happen during a filming shoot. This will need planning in advance. I always make it clear that the process of reaching the nest sites, processing the chicks and returning them back to the nest must happen without any pausing or interference with our normal procedures. The cameras and presenters have to capture what happens there and then. If there is

Figure 7.9 Presenter Iolo Williams doing a live broadcast about Bath's Peregrines on the BBC's *Springwatch* programme.

more than one chick in the nest, then the camera gets a few chances to capture different angles and shots over the duration. Those with the Schedule 1 Licence and managing the activity are in charge, and must ensure that everything happens as it would if the film crew were absent. If in doubt, seek advice or reassurance from a ringing licensing team such as the BTO in the UK. If during filming something was said or filmed that was unsatisfactory and could be misinterpreted if broadcast, then this needs to be made very clear to the production team so it is edited out or explained by the presenter. I have worked only with the BBC for Peregrine-related filming, and protocol has always been strict and adhered to in a professional manner. But this may not always be the case with all production companies. Additionally, the details of a site need to be handled carefully. Some sites may need to be kept confidential, depending where in the country or world the nest is.

Partnership working and collaborating with others is vital for the success of promoting Peregrines, and in most cases it works very well. Urban Peregrine ringing and promotion needs to bring together everyone who is involved, from the people who own or manage the building to visitors such as church congregations. It is with their cooperation, understanding and facilitation that nest boxes, ringing, filming and special watch points take place successfully. Sometimes, more than one organisation may be involved with promoting Peregrines at a single site – again it is important that all the organisations talk to each other on how publicity or press releases will be handled. This will then lead to positive relationships, transparency and good relations that benefit both them and the Peregrines. If organisations are keen to promote their own work and membership, then an agreement on how this will work to mutually benefit each organisation will lead to a more positive and fruitful engagement, both for those facilitating the activities and those coming to watch the Peregrines.

What does Peregrine ringing tell us about their movements?

The data we glean from both colour-ringed and satellite-tagged Peregrines moving from northern Europe and South America reveals a wealth of information both about their migratory routes and their wintering grounds. It also illustrates how Peregrines may switch between being urban and non-urban birds depending on the time of the year and where they breed.

Urban Peregrines that are not strongly migratory appear to stay on their territory, although passing 'satellite' birds may join them before being chased off. Once fledged, young Peregrines will leave their parents' territory

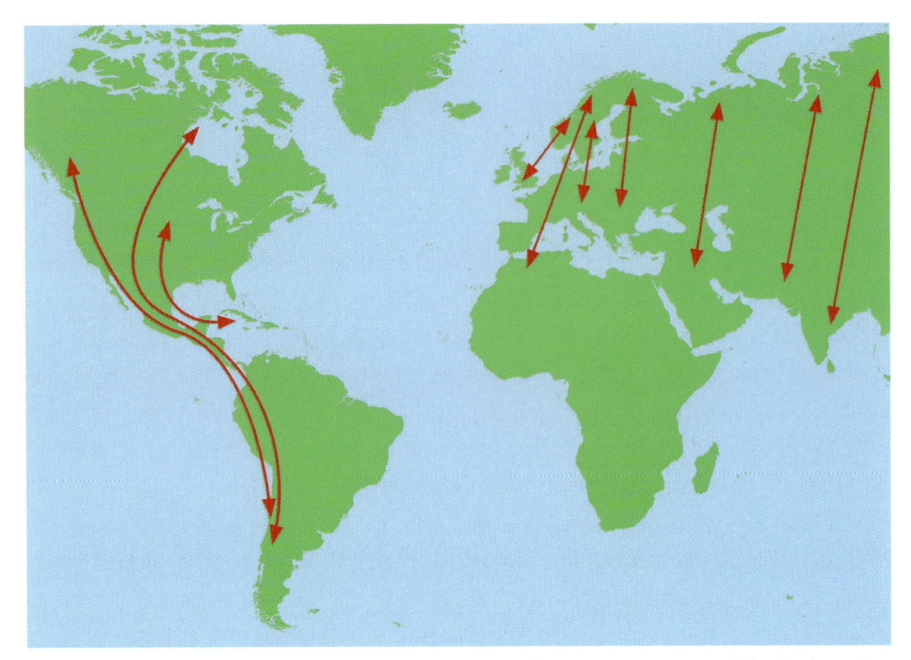

Figure 7.10 The autumn migratory movements of Peregrines across North America and Eurasia.

and move around the region exploring other territories, meeting other Peregrines and honing their hunting skills. Colour-ringing studies from Germany and Sweden show that male Peregrines will stay closer than females to the area where they hatched. In Germany, 40 per cent stay within a 40 km (25 mile) radius, only moving 26.5 km (16 miles) on average; females move 114 km (71 miles) on average from where they hatched. In Scotland, PIT-tagged Peregrines studied by the Lothian and Borders Raptor Study Group showed a similar difference between the sexes, with males dispersing an average of 48 km (30 miles) and females 80 km (50 miles).

Across a lot of the Peregrine's range, the species is strongly migratory, moving large distances between summer and winter sites. Five out of the 19 subspecies, 2 in North America and 3 across Europe and Asia (Fig. 7.10), move particularly long distances. In the UK, Peregrines are less migratory, but they are still very nomadic birds, moving around the countryside and visiting many places along the way.

In Europe they are also very nomadic birds. Populations of Peregrines living in remote parts of northern Europe move south during the winter, many heading to towns and cities in western and central Europe. Those nesting in the very north, such as Finnish and Swedish Lapland, will move

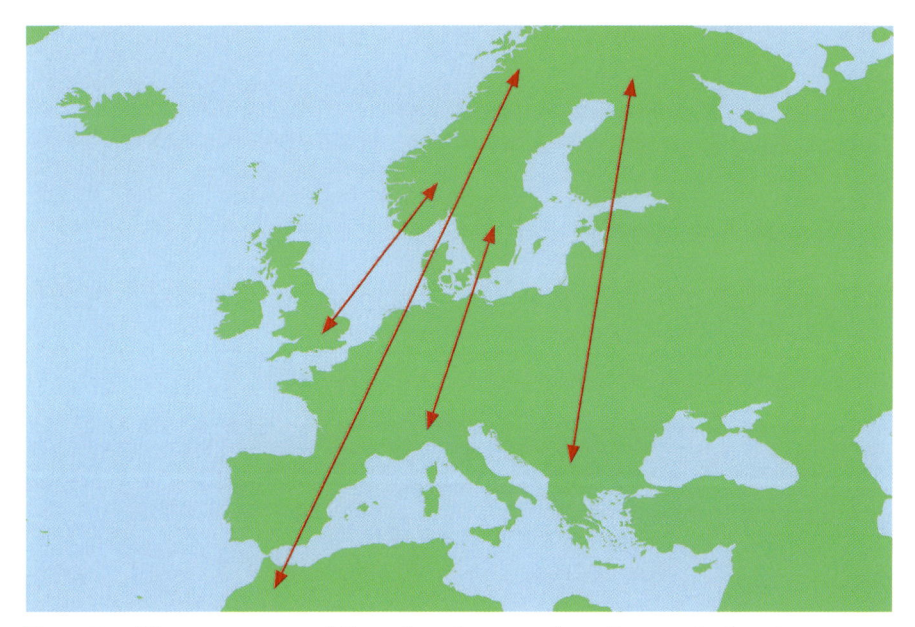

Figure 7.11 The movements of Peregrines from northern Europe during the autumn. Those which breed in the very north leapfrog over those breeding and wintering at a more southerly latitude, and may winter as far south as northern Africa.

further south in Europe and north Africa than those which breed in more southern parts of Scandinavia (Fig. 7.11). Across Siberia, Peregrines also travel south in the winter, taking a wide range of routes, avoiding sea crossings where possible, and wintering in a multitude of habitats and destinations. This helps protect the species from environmental problems. If a disaster happens on one part of its wintering grounds, only a proportion of the Peregrine population may be affected, unless it is a widespread problem such as the pesticide DDT (dichlorodiphenyltrichloroethane) has been in the past. Between 2009 and 2012, Peregrine researchers Andrew Dixon, Aleksandr Sokolov and Vasiliy Sokolov satellite tracked Russian Arctic Peregrines from their breeding grounds in the north to their wintering grounds further south in the Middle East and Asia. Two of these birds decided to winter in urban areas, namely Baghdad, the capital city of Iraq, and New Delhi, the capital city of India. The male bird in Iraq was found hunting over an area of 12 sq km and favoured some of the taller buildings and cranes in the city, such as the Baghdad Tower and the Al Rahman mosque. The female Peregrine wintering in New Delhi was more mobile during her time there, and probably favoured a greater variety of habitats to feed in nearby.

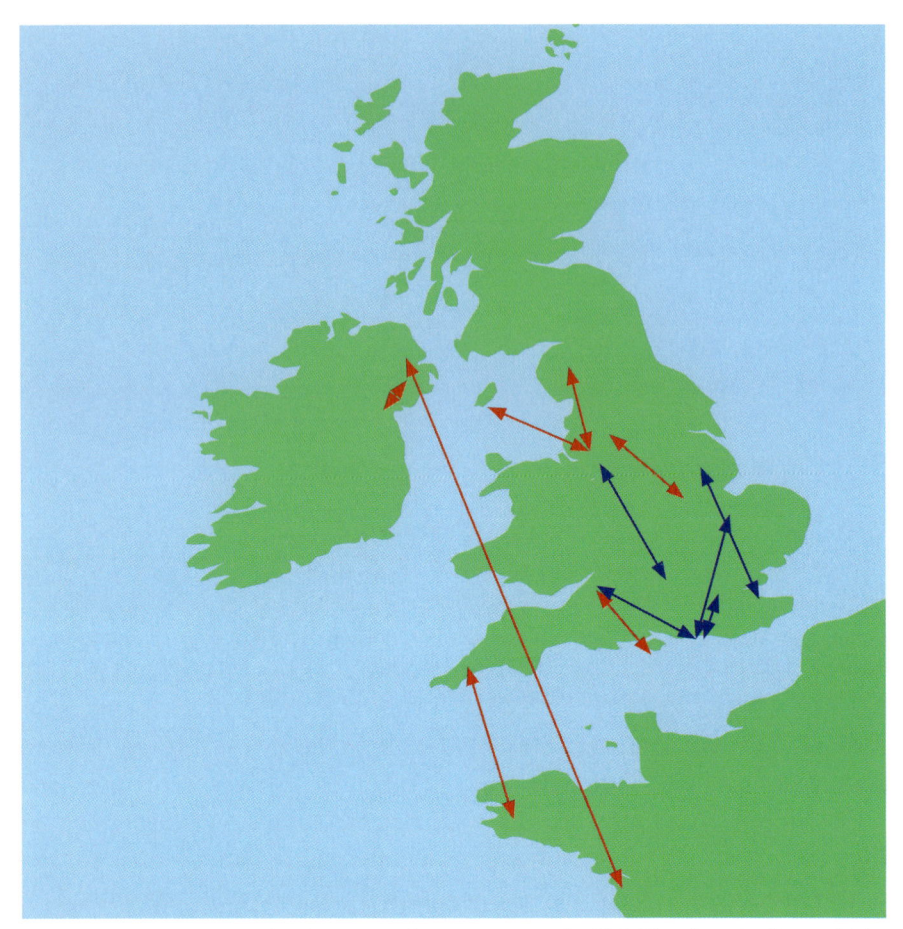

Figure 7.12 Movements of colour-ringed Peregrines in the UK. Blue lines indicate chicks ringed at an urban nest location; those in red were ringed at a rural location.

The UK's wet and relatively warm climate means Peregrines here don't have to move south in quite the same way, and those which hold territories may stay in them year-round. In Scotland, for example, Peregrines may simply move to the lower ground in their territory. In Hungary this behaviour happens a lot and is known as vertical migration – Peregrines move down away from their mountain territories to winter on the lower ground and open plains. Peregrines colour-ringed in the UK generally stay in the country, although one ringed in County Antrim in Northern Ireland in 2005 was spotted in 2013 in France on the edge of the Bay of Biscay, 650 miles away. A Cornish-ringed Peregrine also made it to France, and another to the Netherlands. Generally urban Peregrines in the UK head out in a variety of directions, although those in the south may travel along the coastline and north, rather than heading across the English Channel (Fig. 7.12). Recent ring recoveries show that some may move only a few kilometres or even stay where they hatched, while others will head anywhere between 123 and 213 km (76 and 132 miles) away.

In North America, Peregrines move south for the winter, heading for a warmer climate in Central and South America (Fig. 7.10). The Southern Cross Peregrine Project in the USA has been tracking Peregrines from their wintering grounds in Chile back to their breeding areas in the Arctic. The journey may take up to 48 days, clocking up over 14,271 km (8,868 miles). Satellite tracking reveals that the Peregrines will travel up to 602 km (374 miles) across the ice of Hudson Bay, Canada, before it has melted in the spring. One individual, known as Island Girl, left her winter site in mid-April in 2013 and arrived back north-east of Hudson Bay at the beginning of June. She has repeated the same journey over similar dates since 2009. The journey reveals the variety of habitats she passes through, from her starting point on the coast in Chile, through some of the driest desert landscapes on Earth and on through tropical jungles, mangroves, low coastal plains, agricultural fields, rivers, shrimp farms, mudflats, rocky canyons, and urban areas such as a shipping terminal. Stopping on the way she may rest on pylons and telecommunications towers. Her journey takes her through places that would make a backpacker envious, including Peru, Ecuador, Colombia, Mexico, and through the USA towards Canada. The studies reveal that Peregrines such as Island Girl are spending only five months on their winter grounds, two months each way on migration, and three months at their breeding sites, which may be covered in snow and frozen on arrival – a contrast to Peregrines further south, which will be fledging chicks at the beginning of June. Some Peregrines on their return journey south from the Arctic may make it to Europe, probably by accident. The journey south may be treacherous, and Peregrines need to avoid storms and tornadoes. On the way they may pass resident urban Peregrines, for example in the cities of Lima, Peru and Cali, Colombia.

As with Swallows, warblers, shrikes, orioles and ducks, there is a very good reason why migrating north from the wintering grounds offers the Peregrine an advantage. It all comes down to food, less competition with other species, and the long daylight hours. By travelling hundreds or thousands of miles north there is an abundance of food available, together with more light hours during the day and fewer other raptors to compete with. Back in the tropics and lower latitudes the days are shorter, with more competitors all after the same supply of food of small to medium-sized birds.

CHAPTER EIGHT
Myths about Peregrines

THERE ARE ALWAYS MYTHS about the natural world – stories, ideas and interpretations about animals or plants which are passed down generations or between people but which are not true. Often they are assumptions, misinterpretations, misunderstandings or simply theories, because at the time they were thought up the technology or research wasn't available to prove otherwise. In the eighteenth century many naturalists believed Swallows hibernated below the mud of ponds when they disappeared from England in the autumn. And in the Middle Ages, Barnacle Geese *Branta leucopsis* were thought to hatch from goose barnacles. The geese disappeared during the summer and it was thought that they emerged in winter from the feather-like appendages of the crustacean.

Some of the classic myths relating to Peregrines are not anywhere near this extreme, but nonetheless they have been dispelled over time by the advent of solid research, changes in technology, and a better understanding of science.

So let's bust a few myths about Peregrines.

Myth 1: Urban Peregrines just feed on pigeons

Figure 8.1 Feral Pigeons are not the only type of bird urban Peregrines eat (Adam Rogers).

Peregrines don't just feed on pigeons, whether they live in the countryside or in cities and towns (Fig. 8.1). My own research on urban Peregrines reveals that generally only half their diet is pigeons, meaning the other half is made up of a variety of other birds.

There are lots of pigeons in urban

areas, and Peregrines are more often than not depicted with a pigeon in their talons, or in hot pursuit of one. By looking more closely at what the Peregrines are leaving below their feeding sites, we now know there is a lot more to the falcons' menu than just pigeons. They eat everything from gulls to Goldcrests, and Blackbirds to Blackcaps.

Myth 2: The bird killing the pigeons in my garden is a Peregrine

Peregrines don't hunt in gardens and are very unlikely to be ever seen in one. A big female Sparrowhawk feeding on a Collared Dove or a Starling is often mistaken for a Peregrine. However, there are key differences in how the two feed and live. Sparrowhawks hunt by flying in short bursts, dashing through woodlands and gardens, twisting and turning to catch an unsuspecting bird. Once the bird is caught, the Sparrowhawk will take it to a favourite plucking branch, or will pluck and eat it on the ground. Instead, the Peregrine needs wide, open spaces where it can stoop dive to catch prey or fly low over pools and ditches to catch unsuspecting birds. Its long, pointed wings and proportionately shorter tail don't give it the manoeuvrability to dash between the trees. However, this is not say that Peregrines don't on rare occasions appear in gardens, especially if they are young and recently fledged, or are coming to a stream or pool of water to wash. The owner of a large garden near Bath once told me her stream running through the garden was host to a Peregrine which came down to wash from time to time. While there was no photo evidence, Peregrines will come down to water bodies to wash and keep their plumage in good condition. Unlike small birds such as thrushes and sparrows, the Peregrine gets all its water from its food, and is unlikely to be visiting for a drink. Peregrines will sometimes take small birds such as Treecreepers, tits and Goldcrests, so it is also possible that they do occasionally hunt adjacent to or within woodlands, using the element of surprise to catch their prey.

Figure 8.2 A Sparrowhawk like this juvenile is a common garden visitor, unlike a Peregrine. The bright yellow or orange iris is characteristic of a hawk.

To tell Peregrines and Sparrowhawks apart, it is also worth looking at their eyes – Peregrines have dark eyes, while a Sparrowhawk generally has yellow or orange irises (Fig. 8.2).

Myth 3: Is a Peregrine an eagle or a hawk?

The Peregrine is not an eagle, although both are very impressive creatures. An eagle such as a Golden Eagle *Aquila chrysaetos* is twice as long as a Peregrine, with a larger, more powerful beak, huge talons used for grabbing hares, foxes and deer, and long, broad wings. The Peregrine is tiny in comparison, with thin, pointed wings. Both come from the same order of raptors, Falconiformes, but the Peregrine belongs to the Falconidae family and the eagle belongs to the Accipitridae family.

Unlike Peregrines and other falcons, hawks have broad wings and bright yellow or orange eyes. A large female Sparrowhawk may be as large or larger than a male Peregrine, while a male Goshawk may be as small as a female Peregrine. However, Goshawks are often larger and capable of catching and eating a Peregrine. Like eagles, hawks are from the Accipitridae family rather than the Falconidae family.

Myth 4: Rather than catching prey on the first hit, Peregrines give it a glancing blow

Occasionally, Peregrines may hit their prey with a glancing blow, stunning it or perhaps killing it outright on impact, and then flying back to retrieve the falling bird. However, in a stoop dive they generally catch their prey on the first hit.

Myth 5: Peregrines hover

Kestrels are a familiar falcon on motorways and roadside verges (Fig. 8.3). Their typical hovering behaviour is widely known and remarked upon, even by people not interested in birds. Peregrines are often mistaken for them, but unlike Kestrels they generally don't hover. Kestrels have perfected their hovering technique to catch voles and mice in grassland. On a windy day you may see Peregrines holding themselves in the air and almost hovering, but usually without the regular wing beating of a Kestrel. Peregrines may also hover briefly over water to catch surfacing grebes or diving ducks.

Figure 8.3 Kestrels are often mistaken for Peregrines. However, they are smaller and they hover.

Myth 6: Peregrines and Ravens live in harmony

While Peregrines and Ravens generally live side by side, there is sometimes a love–hate relationship (Fig. 8.4). Ravens will sometimes steal the eggs of the Peregrine, and occasionally Peregrines may kill Ravens. But usually Ravens are just harassed and the falcon's stoops come to very little. If there are threats close to the nest of either species then the Ravens are good at raising the alarm and mobbing, while the Peregrines will persistently attack and see off the threat.

Figure 8.4 A young Raven may be harassed by Peregrines but is rarely killed.

Myth 7: Peregrines eat big urban gulls and help control their numbers

Peregrines eat small gulls such as Black-headed Gulls and Common Gulls *Larus canus*, and occasionally they may kill a Herring or Lesser Black-backed Gull (Fig. 8.5). However, they are more likely only to injure the latter two in cities when protecting their territories and chasing them away. Those roofs closest to a Peregrine nest site may be empty of large gulls as the adults are likely to be chased away repeatedly, beaten up, or have their chicks eaten when they fledge. Generally, though, the two groups of birds live side by side, as we see on cliffs on the coastline. In areas where smaller gull species live they feature in the Peregrine's diet more often – their size makes them easier to catch and they don't fight back in the same way a larger gull will.

Figure 8.5 Small gulls such as Common Gulls (in this photo) and Black-headed Gulls are easily caught by Peregrines, while larger Herring and Lesser Black-backed Gulls are rarely killed and eaten.

Myth 8: Peregrines will control a population of pigeons

Peregrines are only one predator of pigeons, and there are many other reasons why pigeons may die or be killed in towns and cities (Fig. 8.6). Additionally, Peregrines don't just eat pigeons; they eat lots of other birds, too. While Peregrines may help keep pigeons in check, they are only one of many contributing factors which control urban pigeon populations. For example, as urban gull populations increase, more and more pigeons are killed and eaten by the larger Herring and Lesser Black-backed Gulls.

Figure 8.6 Pigeons have many pressures on their populations, with predation being just one factor that influences their survival (Adam Rogers).

Chapter Nine

Changing Threats and the Future of the Urban Peregrine

The Peregrine is an incredibly successful, cosmopolitan bird, catholic in its diet, able to respond to the changing seasons by migrating long distances, and making use of a wide range of habitats. However, despite its comeback around the world during the past 30 years or more, we cannot be complacent about the Peregrine's success (Fig. 9.1). In many parts of Europe its population sizes are still relatively low. Even within the UK, despite its success and range expansion, some of its breeding populations in its northern range (northern England and Scotland) are declining. In some places, the Peregrine is at a lower density than it could be or is totally absent. The Peregrine's relative, the Kestrel, was once very common in the UK, but a drive along the motorway or a country lane where Kestrels were once almost guaranteed often results in no sightings today. A bird we had almost taken for granted is suddenly slipping away on our watch. Fortunately lots of research is going on to find out why the Kestrel is declining and hopefully it won't be too late. But the same could happen again with the Peregrine if we don't keep a close eye on it. Over hundreds of years the Peregrine has shown a see-saw in its population, with its numbers going up and down, usually in relation to human activity. Having a better understanding of how the bird lives and deals with different ecological situations can help secure its future.

Raptors have never had it easy – for as long as man has been looking after game birds such as pheasants, there has always been a conflict. There is a misunderstanding by some about the relationship between most raptors and their prey, and changing negative attitudes towards raptors is difficult. Despite the hard research, legal protection and partnership working between interested groups and stakeholders, there are still many birds of prey being

Figure 9.1 While we can't be complacent we are still able to celebrate the return of the Peregrine (Sam Hobson).

needlessly and unacceptably persecuted by a minority in the UK and other countries. Science and common sense continue to be undermined by profits, jobs, bloody-mindedness, and Victorian attitudes.

But let us celebrate the Peregrine. Across many parts of its range, its population is at levels that haven't been seen for hundreds of years. And in those countries where it is still recovering, there is ongoing work, such as captive breeding programmes, to give it a helping hand. For example, between 2010 and 2013 a team in Poland has released 339 Peregrines into the forests to increase the tree-nesting population that went virtually extinct 50 years ago.

Peregrines, like any animal, have many natural causes of death. We are very quick to assume that when a Peregrine dies it has been the result of human interaction. But often the reason for its death may be down to fighting, starvation or disease, while accidental deaths include collisions with buildings, power lines and trains. Some deaths result from other hazards of living in urban areas, such as getting tangled in netting used to protect buildings from pigeons and gulls, or young birds drowning in rivers and bays when they fledge. Those breeding around industrial estates occasionally fall to their death down power station and chemical plant chimneys. Many of these accidental deaths can be mitigated over time. Predation is also a natural threat – depending on where Peregrines are nesting, their eggs, young and the incubating adults can be killed and eaten by large owls, hawks, mustelids (such as weasels and martens), foxes and other predators.

A lot of deaths, however, are related to direct intervention that is illegal, unnecessary, and lacks any understanding or sympathy as to how the Peregrine lives and fits into the ecosystem. The behaviour of those involved may be an obsessive hobby, a way of making money, or the result of an order given by their employer. For others it may be an action resulting from frustration and anger due to conflict with their own hobby or interests. Either way, their actions are illegal, inexcusable and will often go to court. The type of crimes committed against Peregrines include shooting them, poisoning them by leaving food laced with toxic chemicals, destroying nests, and using traps which lead to the birds being fatally injured. In southern England, rural and quarry Peregrines generally appear to be doing well, although poisoning and egg collecting crimes still happen. Further north, especially in remote locations, Peregrines are often absent. Pairs may suddenly disappear, or have their breeding success compromised. The illegal removal of Peregrine chicks from their nests is also a problem. The chicks are used in falconry around the world to inject fresh genes into the captive breeding population. In the UK this is illegal, but in some countries licences allow a small quota of chicks to be removed from wild nests. Despite this, it remains a controversial technique, especially when many populations are still just recovering.

If it is regulated carefully where Peregrines are doing well, legally taking chicks for conservation and falconry purposes may be viable longer term, but would need further scrutiny, discussion, and amendments to laws or licences in those countries.

The law

Peregrines are a Schedule 1 species in the UK, which means that they have special protection under the law to stop them from being disturbed, injured or killed, particularly during the breeding season. Their nest sites are protected during the breeding season (generally mid-February to mid-July) and researchers studying the eggs and chicks are required to have a Schedule 1 Licence, and must visit the birds under strict procedures to minimise disturbance. This protects the Peregrines from illegal activities but also disturbance actions that may be non-intentional. Urban Peregrines may allow people below them to watch at close quarters, but they are still vulnerable to disturbance and activities that may be associated with the maintenance and day-to-day activities of running a building, for example. The law protects both Peregrines in the countryside and those that are living on buildings. In rural locations where Peregrines see fewer people, disturbance distances can be large – Peregrines in Scotland may start showing some signs of concern from a few miles away. Those living in cities are far more resilient and will be relaxed with people minding their own business on pavements below, sometimes only 10 to 20 m away. In the USA, the U.S. Fish and Wildlife Service has removed the Peregrine from the list of threatened and endangered species, but it can be added back any time in the future if there is a need for further protection. The Peregrine is also protected across Europe and many other parts of the world, and like many raptors it is offered some of the highest levels of protection under both national and international laws.

The story from the past

Peregrines are probably at their highest levels in many parts of the world, and certainly in the UK, than they have been for hundreds of years. Back in the 1700s, Peregrines and other raptors had a fair bounty on their heads. Anything perceived to be a threat to animal husbandry or plant harvests was killed. Different amounts were paid as a reward depending on the type of animal.

In the Zoological Notes (p.51–2) in *The Annals of Scottish Natural History* (v14. 53–6) in 1905, the following is written:

> *Scottish Lists of Vermin. In the Annals for July last Mr. Harvie-Brown gives a list of vermin killed on the Glengarry estates, Inverness-shire, between 1837 and 1840, with the remark, I think this list has not appeared before. As a matter of fact it was published in 1850 in Knox's Game-Birds and Wild-Fowl, p.117. Not only so, but Knox's list is fuller, and contains the following species not mentioned in the Annals: Ospreys, 18; Common Buzzards, 285; Bluehawks or Peregrine Falcons, 98; Orange-legged Falcons, 7; Hobbyhawks 11 … I can now give a much older list of vermin killed in Aberdeenshire (1776–1786), which I copied some years ago into one of my note-books from the London Chronicle of 18th September 1812. It runs as follows: Previous to the year 1776 the destruction of sheep by ravenous animals was so great in the wilds of Aberdeenshire as to be computed to be equal in value to half the whole rents; and the game and poultry suffered in full proportion. At this time a subscription was entered into, and applied to premiums for removing the evil. This continued for ten years, when it was found that during that time, in five parishes only, 634 Foxes, 44 Wild Cats, 57 Polecats, 70 Eagles, 2520 Hawks and Kites, and 1347 Ravens were killed, besides many which died by poison, or of their wounds, and sheep have since been in perfect safety at all seasons in those parts.*

This account gives just a brief idea of the type of misguided persecution that happened back then. It is mind-boggling to know that on some game estates today time has stood still and such persecution still continues. Even with all the current research, protection, laws and efforts to find solutions, many of the species mentioned are still killed as if we were living back in the Middle Ages.

During the Victorian period in the UK, animals including Peregrines were shot for taxidermy and displayed in houses and museums. While it may be out of fashion in the UK these days, the technique still continues in other parts of the world and poses a threat to Peregrines in some places. In countries such as Bulgaria and Russia there has been a rise in the desire for taxidermy animals, including Peregrines, while across northern Africa, the Middle East and Asia, Peregrines are at risk from hunting, taxidermy and trapping for falconry.

During the Second World War, Peregrines were purposefully shot to stop them catching pigeons that were carrying important messages from war ships to sites inland. Today, many museums still have drawers containing Peregrines shot as part of the 'war effort' (Fig. 9.2). During this time, and earlier during the First World War, lots of gamekeepers went to war and many never returned. With fewer people in the countryside to kill those animals they labelled as vermin, and with additional protection from the Wildlife and Countryside Act during the twentieth century, the populations of many species were finally able to recover. The Buzzard has been able to spread across every English county, while the Red Kite *Milvus milvus* once more graces the rolling countryside in southern England and other parts of the UK where it has been reintroduced. Peregrines also enjoyed a comeback post-war.

Figure 9.2 Peregrines were shot during the Second World War to stop them intercepting pigeons carrying important messages. Some of those Peregrines survive today as study skins in museums.

But the most devastating threat to Peregrines and other raptors was still to come. During the 1960s and 1970s, farmers used a chemical pesticide known as DDT. Pigeons and other birds feeding on sprayed plants and seeds ingested the chemical, and over time it built up to high levels in the bodies of predators that were eating these birds and often killed them.

Even if DDT didn't kill, it had other catastrophic effects. Peregrines began producing eggs with very thin eggshells, and clutches of eggs were simply crushed under the weight of incubating adult birds. This affected not just Peregrines but also other birds of prey such as Sparrowhawks, Buzzards and Kestrels. The late nature conservationist and Peregrine researcher Derek Ratcliffe discovered the link between DDT and the decline in raptor populations both in Britain and across the world where the chemical was also being applied. It was a revolutionary breakthrough and required detailed detective work. DDT was gradually phased out across the UK, Europe and the USA in the late 1970s and early 1980s, and since then raptor populations have been recovering. However, the pesticide is still used in many other parts of the world, including countries in Africa. As a result, scientists in Sweden are still finding traces of DDT persisting in the environment. It is thought the chemical enters the atmosphere when it is sprayed on to crops in northern Africa. Tiny drops enter the Earth's atmosphere and move north with air currents to Scandinavia, where they fall back to Earth dissolved in rain droplets before entering the soil and hence the food chain.

In countries such as Sweden, Canada and Spain, brominated flame retardants (BFRs) used in the textile and plastics industries have also been detected in Peregrines and are a newer potential threat. Known as Persistent Organic Pollutants (POPs), they are released, for example, into the Baltic Sea at what are regarded as 'safe' levels, where they accumulate in the muddy sludge derived from the water-cleaning processes in the factories and are easily ingested by worms and molluscs. Wading birds such as Spotted Redshanks *Tringa erythropus*, Whimbrel, Redshanks *Tringa totanus* and Dunlin *Calidris alpina* passing through the estuaries on migration then eat the 'polluted' marine food before flying north to the tundra, where Peregrines eat them. Over time there is potential for these chemicals to build up in the bodies of the Peregrines, just like DDT, and interfere with their natural hormones. BFRs are also detected in the eggs of Peregrines, where concentrations may be high. These chemicals love fat, and this is where they tend to build up, in the fatty tissues of the birds and the fatty deposits found in their eggs. Perhaps unsurprisingly, such chemicals also persist in our own bodies (including in breast milk), and we see their detrimental effects on whales and dolphins. Orcas *Orcinus orca*, which are at the top of the food chain in the sea, eat a whole host of fish, including salmon and other smaller predators which have ingested POPs from their prey. Orcas therefore absorb high concentrations of pollutants from their diet. These build up in their fatty tissues, affecting their

Figure 9.3 Bees are thought to not be the only casualties relating to the use of neonicotinoids in our countryside (Liz Shaw).

health in a number of ways, including a lower sperm count and a less effective immune system. These chemicals are also transferred to their offspring, both at the stage when the foetus is developing and later when the young suckles fat-rich milk. When the calves are born they may already have very high and dangerous concentrations of chemicals such as polychlorinated biphenyls (PCBs) in their bodies, which can lead to an early death.

Some of these chemicals, such as polybrominated diphenyl ethers (PBDEs), have already been banned in Europe, while other similar concoctions, such as penta- and octabrominated diphenyl ether, have been banned in North America. Some illegal products persist in items that are still in use or have been thrown away in landfill and leach into the environment. However, there are others being produced that are thought to be safer alternatives and which are more strictly regulated.

Another set of chemicals to keep an eye on is the group of pesticides known as neonicotinoids. Their effects are still little known despite their widespread use across our countryside, and some believe they could be the next DDT disaster. More research is needed into their effect on nature, though it is known they kill bees and other pollinators, while other important insects may also be affected (Fig. 9.3). They are also more toxic than DDT – smaller quantities are not only more effective at killing insects (and other animals such as birds), but the chemicals they contain are water-soluble and have the potential to persist in the environment for a long time. A decline in pollinating insects also means less food for birds, and in turn less food for the Peregrines.

What's the future for urban Peregrines?

The great thing about urban Peregrines is that they are much loved and have a huge following. In many countries around the world they are well known, and collectively they are watched by millions of people each day, both in real life and online. Their day-to-day health and well-being are close to the hearts of those watching, and if anything untoward happens then someone will notice and raise the alarm.

An immediate concern I have for the explosion of Peregrines in our towns and cities is the safety of their chicks when they leave their nests. In a rural environment the young birds can walk along a ledge, or if they fly clumsily they can land on a nearby ledge or bush. However, in an urban environment the first flight often means a large drop to the ground, especially if the bird is still too heavy to fly, doesn't have strong enough wing muscles, or the wing feathers are still to short (Fig. 9.4). If this happens the bird may instantly break a wing or its back, sometimes with fatal consequences. One juvenile Peregrine was even found with one wing completely missing. However, more often the chicks are just grounded and unharmed, but they remain vulnerable to cars, dogs, and concerned members of the public picking them up or calling for help from a wildlife hospital. The crucial thing is for the bird to remain safe, and often a quick hand from those involved with the nest site to get the bird back to the nest or a nearby roof is all that is needed. Assistance from a local ringer or Peregrine researcher with a Schedule 1 Licence is required if the bird is returned to its nest. In Bath in the UK, the police station is always on alert and has a system in place to call individuals involved with the church nest site to assist with any grounded chicks.

Figure 9.4 This young Peregrine became grounded at an airfield. After being checked over it was released the following day and put back on the hangar.

If a chick has already been removed from the site by people, then after a clear bill of health from a vet or wildlife hospital it is imperative that the bird is returned to the nest site, although caution needs to be taken if the nest site is inaccessible and the bird is likely to end up in water, being eaten by a dog, or run over by a car. I have worked closely with the RSPCA's West Hatch Wildlife Centre in Somerset to help ensure that healthy Peregrine chicks are returned as quickly as possible to where they came from. This also avoids the young birds having to spend more time in captivity than they need to,

Figure 9.5 This juvenile Peregrine was brought in to the Hawk Conservancy Trust, Andover, with a missing wing (National Bird of Prey Hospital, Hawk Conservancy Trust).

and means that they don't need to be 'hacked' back into the wild (the process of slowly reintroducing them). However, sometimes it may not be possible to return them safely, for example if they have come from a pylon nest, in which case they would have to go through the hacking process.

To find out more about injured urban Peregrines I visited Ashley Smith, Kim Kirkbride and Hamish Smith at the Hawk Conservancy Trust in Andover, Hampshire – a falconry centre with a wildlife hospital for raptors.

Since Peregrines have become more urban in southern England, the Hawk Conservancy Trust has seen more birds arriving each year injured or simply grounded. Kim works in the National Bird of Prey Hospital at the Hawk Conservancy Trust, and since 2000 they have seen Peregrines suddenly coming in to the centre in small numbers each year. In some years just a few may arrive, while in others there may be up to half a dozen or more. Forty have arrived between 2000 and 2013, and only 15 per cent are adults, the remainder being juveniles and first-year birds. Sixty per cent have been returned back to the wild. Their injuries or reasons for entry into the hospital vary from simply being grounded to having broken wings and backs (Fig. 9.5). Some are just very thin, while others may have been blown out of their nest in strong winds or had their nest collapse. A few have been caught in netting and one was suffering from lead poisoning. A thin, underweight first-year bird found in Dorset had originated from Sweden. These experiences all resonate with very similar instances and numbers from the RSPCA's West Hatch Wildlife Centre, where most Peregrines come in as juvenile or first-year grounded birds, have wing or back injuries, or have fallen from a collapsed nest. One was covered in mud from falling into the Severn Estuary, and another, a first-year bird found on the Somerset Levels with a leg injury, had flown there all the way from southern Sweden.

I asked Ash what makes Peregrines so attractive to people. Ash feels the Peregrine is a very proud bird and remains distant and wild, even those that have been tamed and fly from the hand. Young Peregrines have their own personalities. Young female chicks are more sensible than their brothers, who tend to be highly strung and, as Ash describes, leave the nest as if they have been fired from a gun (Fig. 9.6). This probably explains how a chick in Bath in 2012 ended up in the river adjacent to the church nest not just once, but twice. And then was never seen again! There is also a tendency for the young males to get henpecked by their sisters. Being smaller, the males no doubt grow their feathers more quickly and are able to leave the nest earlier than the females, who take more time to develop.

Figure 9.6 Young Peregrines, especially males, are very adventurous – this one is checking out the guttering and drainage of a church (Hamish R. Smith).

Ash also talked about some raptors coming out of a 'honeymoon' period in terms of their perception by people, with the hope this doesn't happen to the Peregrine. For example, the Red Kite is an enigmatic bird and one species I always look out for when travelling between London and Bristol (Fig. 9.7). Whether I spot them from the train or while driving on the motorway, their elongated, trailing wings, long forked tail and gliding flight make them instantly distinguishable from the common, broader-winged Buzzard. When I

Figure 9.7 Have Red Kites come out of their honeymoon period?

was nine I remember the RSPB's campaigns to bring back the Red Kite, and throughout my teenage years there was little escaping the awareness of how on the brink the kite was. Twenty years later, the Red Kite has been reintroduced at various locations across the UK and is doing remarkably well. Both the original Welsh population and the reintroduced birds have been breeding in high numbers and spreading – so much so that they are also becoming urban birds. Sitting on a train at Reading or Didcot Parkway, you don't have to look far to spot kites gliding low over the houses. I have even seen one over Westminster Abbey. Some residents put out food for the kites and enjoy watching them as they swoop down to clutch the prey in their relatively small talons. But for some, this is the end of the honeymoon period for the Red Kite. There is a feeling that perhaps they are everywhere, and a definite complacency about their presence in some counties of England. I have often heard people remark on how common they are now, and almost dismiss them as something to get excited about. Fortunately Red Kites are mainly scavengers, taking road kill and other carrion rather than live animals. The Buzzard has been less lucky. After its spread across England over the past 20 years, there have been calls for the Buzzard to be culled in some places due to its supposed persistent eating of young Pheasants on certain shooting estates. Is their honeymoon period over? Well, for many people it most certainly is not, and a huge public outcry in 2012 saw the potential for the Buzzard to be shot on a wider scale scrubbed. However, the following year it emerged that a licence or two had been granted to kill a few persistent Buzzard offenders.

So what about the Peregrine? Firstly, we need to remember that the species almost went extinct in the UK at one time. And in parts of Europe and North America this certainly did happen. The honeymoon period for the Peregrine is certainly not yet over – and perhaps it doesn't exist. While there have been some calls by those who race pigeons to bring the Peregrine down in number, the Peregrine population still has a long way to go to fully recover, and is currently seeing declines and local extinctions in some areas of northern Britain.

London

London feels like a different place for Peregrines compared to other locations, and parallels can be drawn with other large cities such as New York. As you fly in to London, despite its sprawl it is obvious from a Peregrine's point of view where the hotspots are – flying over Stratford, Westminster

Figure 9.8 London at dawn – an opportunity for Peregrines to watch for nocturnal migrants flying across the city (Sam Hobson).

Figure 9.9 London is a mosaic of habitats, from water bodies to parks. The variety of buildings provides the perfect home for Peregrines (David Lindo).

and the City of London, the tall office and residential buildings stand out clearly above the small houses (Fig. 9.8 and 9.9). The coast isn't far as the crow flies, while various water bodies and green open spaces provide a mosaic of habitats and opportunities amongst the railways, roads, and the business centre of Canary Wharf. Common Terns, Ring-necked Parakeets, Feral Pigeons and Starlings provide a constant supply of food for the Peregrines, too.

The London Peregrine Partnership helps monitor and give advice about Peregrines in Greater London, where over 24 pairs of Peregrines are thought to breed, a similar number to New York City but across double the area. Here Peregrines choose to nest not just on office blocks or cathedral/church-type buildings, but also residential flats. The challenges for these Peregrines are very different to those for birds living on other types of buildings. If Peregrines decide to nest on flats, conflicts may arise when the roof or nesting area becomes inaccessible, maintenance work is required, or people who are paying a lot of money for their flat are inconvenienced (such as by a lack of window cleaning). For London Peregrines, the landscape is changing all the time. Buildings get demolished, residents decide they don't want Peregrines nesting, and people access roofs without realising they are disturbing the birds, especially if the nest site is new and unknown. Abandoned buildings with breeding Peregrines in large cities such as London are also vulnerable to vandals breaking in, people exploring the building and others entering to steal scrap metal. For churches and cathedrals, long-term collaborations can be built with local Peregrine researchers and bird clubs to ensure that the Peregrines are able to nest safely and in a way which benefits the venue, stakeholders and the public. On office blocks a similar set-up can be arranged, although further work may be required to ensure that there are solutions with regard to maintenance, access and procedures. Those helping to secure nesting sites for Peregrines on residential blocks of flats may need to liaise closely with residents, building managers and committees. Volunteers from the London Peregrine Partnership help to mitigate problems and provide advice and solutions when these things occur and are known about.

Predator–prey dynamics

Peregrines eat mainly birds and therefore may conflict with the interests of people rearing birds as a sport or hobby. This may include pheasant rearing and pigeon racing. There is also concern from some in the UK that the increase in raptors is directly linked to declines in songbirds.

Extensive research does show that while pigeons are eaten by Peregrines, huge numbers also fall victim to other hazards, from hitting electricity wires to drowning. Meanwhile, the decline of specific songbirds is complicated – research reveals that changes in the environment, food supply, nest availability, seasons and the landscape are all causing changes in bird populations. While raptors such as the Peregrine do eat many types of songbird, the non-pigeon part of their diet is varied and doesn't focus on any one species but on dozens of different types. Therefore there is no significant pressure on one particular species.

Small birds in particular produce a surplus of young to compensate for loss to predators as well as other factors such as disease and weather. For example, as long as a pair of Blackbirds can rear one chick each to replace themselves when they die, then the population remains stable (Fig. 9.10). If they rear more than this then the Blackbird population increases. So whilst a typical garden may see Magpies and other predators (including Hedgehogs *Erinaceus europaeus*, Grey Squirrels *Sciurus carolinensis* and woodpeckers) devouring a nest of eggs or young for the second or third time that season, as long as the Blackbirds have one successful nest (perhaps in the same garden or further away) they have achieved their goal.

Figure 9.10 Blackbirds compensate for predation, disease and weather by having two or three broods a year and producing lots of young.

Figure 9.11 Corncrake feathers found below urban Peregrine sites.

This sounds quite simple, but the problems develop when many different factors begin to chip away at a population of a species to the point where predation and disease can be the final nail in the coffin. The Wood Warbler *Phylloscopus sibilatrix*, for example, is a beautiful lemon yellow-green bird, smaller than a sparrow, which spends the summer in damp oak woodland in the west of the UK. It has seen a dramatic decline in recent years, alongside other trans-Saharan migrants such as the Common Redstart *Phoenicurus phoenicurus*, Pied Flycatcher *Ficedula hypoleuca* and Spotted Flycatcher *Muscicapa striata*. Ongoing research is investigating why these birds are disappearing at such an alarming rate. However, it seems clear that the remaining pairs returning to breed each spring are suffering huge predation pressure from other birds, small mammals and even snakes. Fifty years ago the population of this species would have been able to sustain this level of predation because there were so many more Wood Warblers. But now there are fewer, the effect from predators is greater – but this doesn't mean the predators are the actual original cause of their decline. It is these original causes which need to be clarified and researched to save species such as this.

The same applies to Peregrines and other raptors. While Peregrines may eat the odd rare bird, the reason for its scarcity in the first place is likely to be related to something entirely different. For example, as Corncrakes have started to recover in parts of the UK, they have also appeared more regularly in the diet of urban Peregrines (Fig. 9.11). Not in big numbers, just the odd one each year at a few sites such as Exeter. The Corncrake is reliant on good habitat to nest in, and this is still available in the Western Isles of Scotland. If the habitat disappears there, then the Corncrake disappears, whether or not a Peregrine eats one or two. The appearance of Corncrake in the diet of Peregrines may be a positive sign that they are doing well in some places and are at a level where there are enough of them to begin appearing in the Peregrine's diet during their migration period. Thousands more Corncrakes are sadly trapped and shot in countries that they migrate through. It puts things into perspective that while the odd Peregrine may eat one, we as humans shoot hundreds if not thousands for the pot.

There are occasions when some birds of prey, such as Kestrels, find a colony of birds such as terns and continually target them for food. But again, many colonies of terns or waders are now so fragmented and vulnerable that when a bird of prey is flying across a uniform landscape, a nature reserve with a breeding colony of Little Terns or Lapwings stands out and provides a glut of food. Previously when these colonies were more common and widespread, predation would have had a minimal effect.

Peregrines will only do well if they have plenty of food themselves. There are lots of pigeons in towns and cities, but also plenty of other birds too. If the Peregrines were to cause their prey species to decline, then they too would see their numbers drop, as they wouldn't have anything to eat. In the Valleys of South Wales, the number of racing pigeons available to Peregrines has declined, and therefore so too has the number of chicks the falcons are rearing each year. With less food available, fewer young are able to survive. It is thanks to there being plenty of other birds in the environment that Peregrines are doing well overall. While we may hear of various species in decline, many others, including many that the Peregrine eats, are doing well.

Wind turbines and power lines

Wind turbines have become commonplace in the countryside across the UK and Europe in recent years (Fig. 9.12). Despite their ability to provide important renewable energy, their positioning and aesthetics cause alarm and dismay to some people. For others they are a feat of engineering and add value to the economy, environment and the countryside. Whatever you feel about them, wind turbines can, if positioned in certain places, cause the death of birds such as Peregrines and of other animals such as bats. However, if positioned with good scientific research, background information and foresight, wind turbines can have minimal impact on wildlife. I was involved with the positioning of three wind turbines at Avonmouth Docks near

Figure 9.12 Wind turbines are a potential threat for some Peregrines, particularly on a murky or misty day.

Figure 9.13 This immature male Peregrine died after colliding with power lines, a common cause of death (Jason Kernohan).

Bristol, and aside from a few individual birds being hit, the wind turbines have not affected the birdlife living both on the foreshore and around the docks. Indeed, Ringed Plovers nest right below the wind turbines.

One of the biggest concerns relating to birds and wind turbines is their threat to raptors that may be passing through or hunting over an area where the wind turbines are located. We know, for example, that vultures generally have vision best suited for looking downwards for dead animals, so as they glide along horizontally they may not see a wind turbine in front of them before it is too late. While Peregrines do sometimes hit wind turbines, it is generally thought that they avoid them, and certainly adults may actively keep away from them. However, inexperienced and inquisitive juvenile Peregrines may investigate them more closely. A collision may also occur while a Peregrine is chasing potential prey or being chased by a mobbing corvid. Much of this is all about possibilities, and in reality urban Peregrines don't appear to be affected by wind turbines. As more turbines appear around the world and research is carried out into their effect on birds and other wildlife, we should continue to develop a clearer picture of good practice and the ideal locations which provide lots of electricity but have the least impact on wildlife, including Peregrines.

Power lines also cause a problem for Peregrines, and many are killed as they collide with the thick cables or get electrocuted if they touch more than one cable. Many colour-ringed Peregrines in the UK are found dead after such an accident, and currently this may pose more of a threat to Peregrines than wind turbines (Fig. 9.13).

Hybrids

Perhaps a more surprising threat to Peregrines is the occasional presence of a hybrid falcon in the wild. Such individuals turn up from time to time and are a threat to the pure wild genes of the Peregrine. In many cases permission and licensing is sought to trap or kill the hybrid bird, but some go unnoticed. It is common practice in falconry to breed a Peregrine with another falcon such as a Lanner or Saker. The resulting offspring provide the best of both species, such as speed and size. Occasionally they escape from their owners while out flying and go on their own road trip across the countryside. In 2008 BirdLife International issued a statement asking for a ban on the breeding and keeping of hybrid falcons in the EU as the risk of mixing of genes with wild populations was unacceptably high. As well as Peregrines, other falcons used in hybridisation are also at risk, such as the globally threatened Saker Falcon. The problem is not monitored, so little is known about how at risk the Peregrines are – but with plenty of hybrids in captivity and their potential for reproducing with purebred Peregrines there are concerns, particularly in countries where wild Peregrine populations remain relatively low.

Recreation

Urban Peregrines generally get used to human activity, from cranes moving huge cargo to people walking only metres below buildings they are sitting on. Peregrines will even perch on cranes while they are actually manoeuvring objects and moving around. One pair reported in the 2004 *Isle of Wight Bird Report* nested on a crane of a heavy lifting vessel in Southampton dock, and successfully reared two out of their three eggs, despite going on a 32 km boat journey around the Isle of Wight! The female stayed on her eggs apart from a brief flight, while her mate stayed on a building back in Southampton. Despite this, Peregrines are still vulnerable to disturbance, and their counterparts living in more rural settings are easily disturbed by the gradual increase in recreational activities as more people are taking up outdoor sports. In Bulgaria, the Czech Republic and other Eastern European countries, a surge in residents and visitors exploring the mountains and cliffs through climbing activities means sensitive Peregrines not used to seeing people are easily spooked. In California, people dive-bombing off coastal cliffs and rocks cause similar disturbances.

Other disturbances

In 2013, Peregrines nested on a telecommunications tower in Southampton, leading to the mobile phone company who used it to switch off the transmission while the birds were nesting. It meant that for a one-mile radius users were unable to receive a signal, but it did mean that the birds would be protected and undisturbed by maintenance staff needing access to the structure. Many Peregrines also nest on pylons, and similar situations arise where technicians need access to the structures and the cables. Work has to fit in with the breeding season of the Peregrine, or special permission has to be granted under licence to allow access and minimal disturbance.

Chapter Ten

People and Peregrines

People and Peregrines have lived and worked together for hundreds of years. From watching and admiring wild Peregrines to using them in falconry and honouring them as a status symbol, people have always regarded them as a high-profile bird. *The Peregrine*, a well-read book written by J.A. Baker and published in 1967, romanticises our relationship with the wild Peregrine. The author's journal of events while watching rural Peregrines connects the reader to a full-time occupation of watching, observing and obsessing over a single species. It also provides an escape for the reader, and there is a desire to replenish their own busy life with those experiences described by the author. Today, the name Peregrine and the bird's image are used for all types of businesses, from diamond mining to financial services, communications to pubs, and corporate services to a global immigration project. While they may not be linked directly to the bird itself, the name and image conjure up the speed, vigour and power that a Peregrine possesses.

In recent times, public viewing of Peregrines has become a popular pastime for people around the world, and 'Peregrine Watches' have been set up in towns and cities, capturing the imagination of passers-by who have never seen a Peregrine before and didn't realise they lived in cities. Many observers become very attached to 'their' local Peregrines, feel some sense of ownership, and spend many hours watching the falcons.

Although they are generally increasing in numbers, urban Peregrines are still vulnerable to persecution and changes in their environment. The viewing of Peregrine nest sites in urban areas encourages stalwart vigils to watch breeding falcons and ensure their safety and well-being. In many locations, the Peregrines become local stars – in the city of Derby in England, a local pub has been named after them and a beer, known as the Peregrine Pale Ale, has been brewed in their name.

Figure 10.1 A word cloud showing the words people used to desribe their thoughts and feelings about Peregrines. The larger, more prominent words were the most used in responses (image created using wordle.net).

With 80 per cent of people now living in urban areas in the UK, and this trend set to rise, there is an ever-increasing opportunity for people to see Peregrines close to their homes in an urban location – a contrast to 50 years ago when you may have travelled tens or hundreds of miles to see one. Even in countries where Peregrine numbers are still low, more are being sighted in cities during the winter as others come to visit from increasing populations elsewhere. When Derek Ratcliffe wrote his book *The Peregrine* in 1993, there was very little in the way of opportunities for people to see Peregrines. Indeed, Derek talks very little about urban Peregrines. The earlier urban Peregrine sites were still a few years away from being occupied. Almost 20 years later, the advent of web cameras and the Internet, social media such as Twitter and the use of blogs has meant accessing and enjoying Peregrines, even from our own living rooms, has never been easier.

Peregrines are popular with the media, and well liked by people generally. I asked friends and colleagues on Facebook and Twitter what captivates them about Peregrines and what they love about them. I heard back from 27 people, mostly with an interest in wildlife, and a few involved with Peregrine sites directly. The word cloud in Fig. 10.1 reveals the key words that were used most frequently. While not very scientific, it does reveal what pops into people's minds first. The word 'stoop' really stands out, and along with the words 'amazing', 'perfect', 'shape', 'beauty', 'speed', 'time', 'fastest' and 'seeing' there is some commonality in what people associate with Peregrines and what they love about the species. For a lot of people, Peregrines make them feel good – they are in awe of this splendid bird. The attraction, awe and wonder are similar to the feelings many people have towards big cats – the predators' prowess, their forward-facing, human-like eyes, and their sheer ability to kill. Peregrines are cool birds – they are fast, and they ooze vigour, stealth and speed. They have forward-facing eyes (immediately appealing to people), sharp talons, and they kill – all the successful ingredients to become an iconic bird, a species which people devote lots of time to and want to see time and time again. Peregrines, even in an urban environment, provide people with an immediate connection to the natural world.

Why Peregrines? Why are the Goshawk, Hobby or Golden Eagle not met with the same publicity? These predators also have speed, hooked bills and the power to kill. However, unlike these and many other birds of prey, the Peregrine is most likely to be encountered where most people live – in towns and cities. Ospreys and sometimes eagles and hawks also get some attention, but other birds of prey are usually too timid, inaccessible or protected from disturbance to be watched reliably by a large audience. Peregrines are obvious, stand out and are easily spotted. We largely ignore other birds such as Carrion Crows, Blackbirds or Dunnocks, unless a programme by a television company such as the BBC is featuring them. Perhaps it is the Peregrine's rarity that attracts people – although it is relatively common compared to 50 years ago, it is not a bird that appears in high density. It is known for its speed and the fact it kills other birds – this alone captures people's imagination and lures them in. People simply enjoy seeing Peregrines, and there are few other raptors that you can see so easily and reliably in urban areas. In Bristol, for example, you are more likely to see a Peregrine than a Buzzard, Kestrel or Sparrowhawk – the other common raptors found there.

When the Peregrines returned to the Avon Gorge in Bristol in the early 1990s, the Bristol Ornithological Club set up watches throughout the breeding season to both protect the birds and show them to people. Peregrines have been breeding at Symonds Yat Rock in the Forest of Dean, Gloucestershire, since 1982 and have attracted tens of thousands of people during that time. Neither site is in the middle of a town or city, although the Avon Gorge is close to the suburbs of Bristol and only three miles from the city centre. Their success in attracting people is perhaps down to both being very accessible and popular tourist destinations, where people can easily go and see the Peregrines, and where the birds are relatively easy to find and show to people. There are few other birds of prey which will stand out in the open and be as predictable.

Sparrowhawks are a common garden and woodland raptor in the UK, and are seen occasionally in gardens or dash through them daily in the pursuit of prey. Meanwhile, Goshawks are rare and very shy in the UK; Golden Eagles are also rare, and restricted to islands and the Highlands in Scotland. They live in remote places and have huge territories. Hobbies are fast, dashing birds and tend to live out in the countryside – blink and you miss them.

There is an inbuilt attraction that humans have with nature. It is known as biophilia, and it is this need to be connected to the natural world which empowers people to want to be at one with the world and care for and conserve animals and plants. A friend of mine, David Lindo, is a birder in the UK and goes by the name 'The Urban Birder'. David is a very chilled guy and loves finding urban wildlife, particularly birds. He passionately watches birds and promotes his local patch, Wormwood Scrubs in London, which has both its fair share of common resident birds and other more unusual migrants passing through. David's main aim is to get people living in urban places, both in the UK and abroad, to look up and realise there is a whole world of nature on their doorstep. And it is this message that David takes around with him that helps others to realise that an urban jungle is just that – a myriad of buildings, parks and amenities that does contain wildlife. It may not be as rich or as diverse as the suburbs or countryside, but for those unable to get out of the city it can offer a respite, a relief and a joy. Over hundreds of thousands of years we have always been connected to nature, albeit perhaps in an unconscious way. However, as more of us live in towns and cities our environment is becoming a world of concrete, tarmac, vehicles, computers and much less natural surroundings. But we are unable to run away from the fact that we are all naturally biophilic. It is part of us. Even those disinterested in wildlife or nature will have a biological need to be connected with the outdoors one way or another. In the developed world we have become disconnected from this wild side, and connected instead to our smartphones or laptops. Yet across vast parts of the developing world, being close to the natural environment is part of everyday life.

To breathe fresh air, to see plants or to feel the sun on our faces are things we all naturally seek out. In large cities such as Beijing, where 22 million people are crammed in, green parks are a refuge, a place for socialising, and a place to relax. For many, it is a chance to keep fit, to dance to music as a crowd, and importantly to experience the outdoors in a green space. It is away from the traffic, from the tall skyscrapers and from living in a concrete environment.

Since 2009, the RSPB has been providing watch points to show people Peregrines in the UK. Known as 'A Date with Nature', these opportunities are combined with recruiting new members to contribute towards helping wildlife. Between 2009 and 2012, 11 events were delivered each year and the project reached a total of 116,000 people. In Derby, Peregrine watches set up on a Peregrine pair in the city have attracted 14,000 people since 2007, while the webcam has received 2.5 millions hits worldwide. The Hawk and Owl Trust organise a watch point at a relatively new nest site on the cathedral in Norwich, Norfolk. A remarkable 30,000 people visited the nesting pair in 2013, and almost a million people visited the web camera online (Fig. 10.2).

However, we have to be careful things don't turn in the other direction, as they have for foxes, gulls and other wildlife which are often perceived as being 'everywhere' and are seen as pests. Foxes are common in urban areas, and in many towns

Figure 10.2 Watching Peregrines in towns and cities such as Norwich (top) and Westminster, London (bottom) is a big hit with both passers-by and those with a greater interest in Peregrines (top: David Gittens, Hawk and Owl Trust).

and cities are very tame, some even taking food from the hand or entering homes (Fig. 10.3). The rare incidents of foxes causing injury receive wide media attention, and knee-jerk reactions range from culling foxes to making uninformed judgments on how much of a pest they are. Suddenly, the presence of a wild animal so close to our own lives becomes a problem. This is despite the fact that dogs, which we have at home as pets, cause over 6,000 injuries that require treatment per year in the UK, and occasionally even cause death. You are more likely to be hit by a car, stung by a wasp or mauled by a dog than you are to be bitten or nipped by a fox. And additionally,

Figure 10.3 Urban foxes can be very tame in towns and cities. This young fox in a garden in Bristol is very relaxed and approachable even in the presence of the photographer (Paul Williams).

culling has been shown to not work, as demonstrated by previous supposed pest control in a London Borough in the 1940s that had no impact on the fox population. Long-term research by Professor Stephen Harris and his team at the University of Bristol over the past 40 years has helped to dispel some of the assumptions and myths about urban foxes. Despite the perceived increase in this species, many populations have in fact declined in Bristol or have been recovering from sarcoptic mange over the past 20 years. And their overall population has changed very little since the 1930s. The notion that foxes are eating anything found in bins and on streets is also a myth – while they may be seen rummaging in bins (although less so these days with wheelie bins), research at the University of Bristol reveals that a fox's diet is mainly a very natural one consisting of invertebrates and small vertebrates. Additionally, outdoor pets such as rabbits and ducks are at little risk of being eaten as long as they are kept in secure cages or enclosures. We are the ones who make foxes tame and tempt them in to our houses. Attracting foxes to our gardens can still be done, but responsibly.

We have urbanised the world we and other animals live in, and as soon as something comes too close for comfort we feel a need to repel it, to untame it and to protect our property or ourselves. If an animal becomes very common, our reflex is often to interpret the situation as a problem. Starlings and House Sparrows, now uncommon birds in many parts of the UK, were considered pests 50 years ago when they came into gardens and hoovered up all the food left out for other birds. We also decide that certain characteristics make a species a pest – for example, noise, faeces, damage to property, and physical interaction with people or pets.

The Peregrine's increase in urban and rural areas doesn't resonate with everybody. Some call for Peregrines to be culled in areas where there may be conflict, for example with racing pigeons. The attitude towards an increase in a raptor species such as the Peregrine can be likened to the situation with the urban fox. Many people love the species, but there are also many who feel frustrated and angry at certain aspects of its existence. As with the fox, there are assumptions that need to be challenged. Some assume Peregrines are everywhere and exaggerate their numbers, when in fact some populations in the UK are declining or absent. Peregrines get mistaken for Sparrowhawks, and so there is often a mix-up between which of the two species is doing what. We know the majority of pigeons eaten by urban Peregrines are Feral Pigeons (and not racers), and only half the diet, often less, will be pigeon. This proportion may differ depending on the habitat and the whereabouts of racing pigeons and races. Additionally, if we release domesticated birds in large numbers into the environment, we may have to accept that natural predators such as Peregrines will target them.

As birds of prey have increased dramatically, small bird populations have gone down, and there is an assumption by some that the two must be directly linked.

Even some recent declines in Sparrowhawks are blamed on the increase in Peregrines, Goshawks and Buzzards. It is understandable to see how people come to this conclusion, but they are mistaken when it comes to the reality; what the science and the research actually shows. While birds of prey may eat other birds, the complicated food webs that exist and the ability of prey species to produce many young means millions of years of evolution have helped deal with the effects of predation. And while predation may suppress population numbers of birds to some degree, this is vital to keep bird populations in check and sustainable. The problems arise when unnatural changes in the environment, such as habitat and climate change, also begin to take their toll and deplete bird populations. Either way, questions are still raised as to why the populations of raptors are not controlled. In reality, we are looking at this totally the wrong way round, and need to look at our own actions and behaviours rather than those of raptors. We have to keep things in perspective. An increase in birds of prey doesn't automatically mean a decrease in songbirds. Indeed, the opposite is often true – an increase in bird-eating raptors may mean there are plenty of songbirds available for them to eat. And the reasons Sparrowhawk populations are levelling off or declining in some places may be related to other effects on songbirds that leave the hawks with little or no food.

The Racing Pigeon

While both domestic and feral pigeons are the same species, *Columba livia*, the Feral Pigeon is the true street pigeon, living a life on buildings, paths and bridges (Fig. 10.4). In contrast, the racing pigeon is a fine example of a bird bred for speed, reliability and a sense for direction. It is also a bird rather

Figure 10.4 The majority of pigeons eaten by urban Peregrines in towns and cities are Feral Pigeons rather than racing pigeons. However, this will vary depending on the location of the Peregrines in relation to pigeon races and lofts (Adam Rogers).

liked by Peregrines as food. But the Peregrine is only one small cog in a very big machine. Over the years there has been a conflict between the rise of raptors and the demise of racing pigeons. Ironically, it was the frustrated calls of pigeon fanciers for a control on Peregrines after the Second World War which led to a detailed survey in 1960–61 and the realisation that Peregrines were in fact in steep decline and on their way to becoming extinct in the UK (rather than the other way round). Peregrine numbers in 1963 had almost halved compared to pre-war numbers. Back then scientists calculated that the number of racing pigeons that could be taken in a year would amount to, at its maximum, only 0.3 per cent of the then 5 million racing pigeons. Not only was the Peregrine disappearing from the British countryside, it was only eating a relatively tiny percentage of the racing pigeon population.

Today there are many more Peregrines and fewer racing pigeons, though not that many fewer. Across the whole of the British Isles there are around 42,000 pigeon fanciers, distributed across half a dozen unions and associations (Fig. 10.5). A mind-boggling 1.1 to 1.2 million pigeons are bred every year, and in total there are between 4 and 5 million racing pigeons. Hundreds of races take place each year. Racing pigeons fly hundreds of miles between a release site and their home destination, using physical markers such as road

Figure 10.5 A pigeon fancier waits for his birds to return from a race (Royal Pigeon Racing Association).

networks and water bodies as well as their own internal compass to find their way back. Despite seeing a decline in registered pigeon fanciers, the Royal Pigeon Racing Association (RPRA) is actively attracting young people and ensuring the long-term survival of this sport, which is a way of life for some. It is great to see another channel in which young people are able to engage with nature, the outdoors and a long-established sport. Racing pigeons is a competition for all the family, with monetary rewards. Along with racing pigeons, many people also breed and show fancy pigeons. These come in many varieties. Some may be tall and elongated with a huge puffed-out ruff of feathers on the throat and neck, such as the pigmy pouter; while the humble tumbler pigeons roll backwards and side to side in the air. Many of these fancy pigeons, if given a chance to fly from the coop and exercise, are easy prey for predators, being less well-designed for quick getaways.

If pigeons are released into the environment, then the consequences of exposing them to predators are inevitable. Peregrines are natural predators – and by releasing pigeons into the skies to return to their homes, we are providing Peregrines with an abundance of food and an easy target, especially if the racers are young, inexperienced or tired. Large flocks of pigeons flying directly over the countryside and urban areas don't go unnoticed by birds of prey, and Peregrines and other raptors will target individuals as they pass through. Sparrowhawks particularly target pigeons that are out flying on practice flights closer to their coops. But this has to be put into perspective. If a prize pigeon hasn't returned after a race, there may be a number of reasons why. Many racing pigeons simply get lost or exhausted – you may have seen one loitering around your garden, park or woodland where it has just dropped in from a race. Others get electrocuted from hitting power lines, collide with cars, or drown in the sea or other large water bodies.

When the RPRA was formed in 1896–97, originally as the National Homing Union, raptor numbers were artificially low. Today, an acceptance of natural, higher population levels may require further adaptation and change in the sport, especially if raptor populations remain protected under law. There are calls to control the populations of raptors in some places. It isn't disputed that Peregrines, Sparrowhawks and Goshawks eat racing pigeons (Fig. 10.6 and 10.7), although the extent to which this happens is hotly contested and has been for decades. Unfortunately, the raptors are often made into scapegoats, with a misunderstanding of the complex environment the pigeons are living in.

Either way, measures can be implemented during the racing seasons to

Figure 10.6 A juvenile Peregrine with a freshly killed pigeon (Sam Hobson).

reduce the impact on racing birds. These include changing the timing and route of the racing pigeons. A study in the Rhondda Valley in South Wales has shown that not only does this reduce the number of pigeons being taken by Peregrines, it also reduces the Peregrines' productivity, suggesting their numbers along the valley had been kept artificially high when racing pigeons did fly through.

Back in towns and cities, the picture can vary. Researchers at some urban sites record very few racing pigeons in the Peregrine prey remains each year – certainly only small numbers of racing rings or wing-stamped feathers are found. Racing pigeons may account for only 1 to 2 per cent of prey found at any one site. At other sites, perhaps where routes do overfly urban Peregrines, this percentage may be higher, at up to 17 per cent, and it will vary across the UK depending on where a greater proportion of races are taking place and on the concentration of pigeon fanciers exercising and racing their pigeons.

Figure 10.7 A Peregrine pellet containing a racing pigeon leg ring.

Figure 10.8 A Feral Pigeon on the lookout for easy pickings (Adam Rogers).

Meanwhile, wild Feral Pigeons are numerous and a study by John Tully has shown that in 100 years, the population of Feral Pigeons in the Bristol region in western England has changed very little. The pigeons may have altered their favourite places to hang out, but their numbers remain the same. Towns and cities are home to hundreds and thousands of pigeons, and even with more Peregrines now visiting and breeding in these areas there are plenty of pigeons to go round (Fig. 10.8). It is important to remember, though, that half of the Peregrine's diet in urban areas is not pigeons but other birdlife, from Blackbirds to Dunlin and terns to Woodcock.

Other conflicts

I love flying, and whenever I get the chance to have a window seat when travelling across the UK I take it. Looking out across the British Isles from an airplane gives me an opportunity to take in where places are, and how urban and rural areas fit into the wider landscape. We might have digital maps on our computers, but nothing beats seeing it for real. Leaving behind the pretty, patchwork quilt-patterned landscape in the south and heading north above the hills and mountains, these journeys provide a fascinating opportunity to see the places where different Peregrines live (Fig. 10.9). They are a good chance to spot urban areas and see how built-up areas sprawl into the countryside. Over the remote and extensive open country of northern England and Scotland, it is easy to see why Peregrines have much larger territories than in towns and cities. Prey is more widely distributed across the habitat, and Peregrines need more space to find it. However, despite the tranquillity and quietness of these places they are also where more sinister events unfold that are causing entire gaps

Figure 10.9 A view from a plane looking towards the southern edge of the Lake District – a Peregrine's-eye-view of one of the wilder parts of the UK.

Figure 10.10 The Royal Society for the Protection of Birds (RSPB) recovered this poisoned adult Peregrine and its three-week-old chick from a nest ledge in rural Gloucestershire (Guy Sharrock, RSPB).

in the distribution of Peregrines and other raptors. While Peregrines are doing well in urban areas in the UK, some upland areas managed for grouse simply don't have any Peregrines when they naturally should. The birds are routinely and illegally killed by some landowners and their gamekeepers to provide grouse moors devoid of predators. Historic sites good for breeding Peregrines now lack them. Others appear in territories one year only to suddenly vanish or fail to breed. Many adjoining estates also fail to see any Peregrines or other raptors alive and breeding during the summer months. Compared to those in other habitats, Peregrines living in upland moors are producing half the number eggs and chicks, their productivity stifled by human intervention. As I was researching the conflicts that urban Peregrines face, it became clear that real and illegal threats to them in rural parts of the UK also affect those living in towns and cities, especially as some Peregrines move from urban areas to breed in cliffs

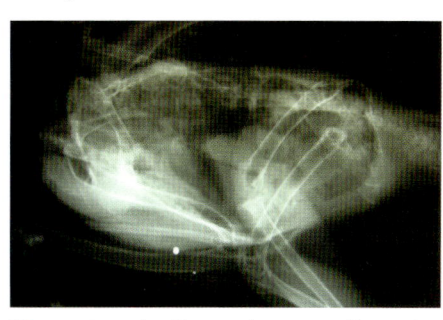

Figure 10.11 An X-ray of a young Peregrine shot in one wing (Vale Vets Ltd).

and quarries in the countryside (Fig. 10.10 and 10.11). Why some landowners do it and how they get away with it can be difficult to understand, but it appears to come to down to money and profits; poor understanding or ignorance of natural ecosystems, predators and research; sheer bloody-mindedness; the remoteness of estates (so perpetrators can get away with it); and contacts (who knows who). It is a real shame, especially when gamekeepers are also the solution to the problem – they know how to best manage grouse moors for grouse and probably know more about the other species that live there than most conservation organisations.

So how can these conflicts be resolved?

Mark Avery, former Conservation Director at the RSPB and now a highly regarded blogger, writer and speaker who stands up for nature, sets out very nicely in his book *Fighting for Birds* three possible solutions for protecting birds such as the Hen Harrier *Circus cyaneus* and Peregrine, which are both highly persecuted on grouse moors. The first is to do nothing and give up! The second is to compromise and look at a positive solution; for example, providing diversionary food for the Hen Harrier so it doesn't catch grouse, or changing the timing and location of racing pigeon routes to avoid the Peregrine. And the third is to campaign against those in the sports that are causing certain birds of prey to be in such a diabolical and unacceptable decline.

Pigeon fanciers love their birds, and they love nature. The majority do not want to see the demise of raptors. However, if a prize pigeon has just been eaten by a raptor it is understandable that its owner will be annoyed, frustrated and want a solution to stop this happening again. For some, the impact from raptors has seen them quit their hobby. But of course this does not excuse taking the matter into their own hands and killing raptors. While this is rare, and most fanciers act responsibly, there are still pigeon fanciers out there with a vendetta against Peregrines. Those that do still destroy Peregrines, sometimes acting in connected networks across countries, are continually being rooted out and prosecuted or dissuaded from continuing.

For me, future resolutions require dialogue between all parties involved. In my opinion, things won't move forward unless there is an understanding of the pigeon fancier's point of view as to what the issues really are. Some problems may well have very simple solutions. Raptors are never going to go away and they are always going to eat other animals. But they will continue to decline if unacceptable and irrational persecution persists

Figure 10.12 The Peregrine is part of a much bigger picture with regard to the urban landscape and the threats that racing pigeons endure (Sam Hobson).

alongside changes in prey abundance or habitat quality. Indeed, hawks may target racing pigeons at a time when other prey is less easily available. They are certainly easier for a Sparrowhawk to seek out and find on a daily basis compared to other birds that move around and hide. We also need to understand the issue from a scientific point of view, but this doesn't mean spurting out figures that only some people understand. It means making good science accessible for everyone to interpret and not feel threatened by. There are many other factors involved in the demise of racing pigeons, including being dispatched by their owners from time to time if they are ill, injured, or have got lost at some point. Controlling raptors is not the solution – but looking further afield at the bigger picture and mitigating for losses to raptors are both key. The Peregrine is part of our natural heritage, our countryside (Fig. 10.12). The racing pigeon is part of our social heritage and social history. With cooperation, partnerships, and informed changes in flight times and routes, a better solution can be sought to reduce the impact of raptors on racing pigeons. We need to deal with and remove the barriers, stigmas, misinterpretations, myths and assumptions.

After hearing so much about racing pigeons and racing pigeon fanciers, I was keen to find out what was really going on with regard to longer-term solutions to the conflict between raptors and pigeons. I wanted to understand

Figure 10.13 A racing pigeon returns to its loft (Royal Pigeon Racing Association).

a way forward which would allow informed decisions to be made to protect both racing pigeons and raptors. The conflicts between raptors and activities such as game shooting and pigeon racing all feel very messy. Whatever solutions are provided by those for or against the arguments, are usually shot down (excuse the pun) and no meaningful steps forward seem to be made. There needs to be sensible dialogue between all parties and we need to stop going round in the same circles decade after decade. There is no dispute that raptors eat pigeons (or grouse), and even how many they take can be calculated relatively easily with both maximum and minimum estimates. But the fundamental question to explore is how do we learn to accept and live with them?

I have been reluctant in the past to discuss racing pigeons in huge detail – mainly because I wanted to be better informed about what I am asking and talking about. Working out exactly where thousands of pigeons which don't return to lofts each year go isn't easy. My own research on urban Peregrine food has shown that in cities such as Bath, racing pigeons make up as little as 2 per cent of the prey species found. And observations by Peregrine researcher Andrew Dixon have shown that there is a bias towards larger prey items such as racing pigeons being discovered and identified at a Peregrine eyrie compared to what is actually delivered throughout the day. Alongside the racing pigeons there are lots of other prey items brought in which may not be found by researchers. Therefore, the percentage of large prey items such as racers can in fact be halved to compensate for the bias, providing a much lower value than may be originally thought or reported.

I decided to meet with Stewart Waldrop, General Manager for the Royal Pigeon Racing Association and responsible for 28,000 loft members in the UK, to find out more about raptors and pigeons (Fig. 10.13). Driving up the M5 from Bristol to Cheltenham I was a little nervous, but keen to find out more about the organisation and positive solutions to their conflict with raptors. Arriving at the association's HQ, a historic yellow Edwardian House, I gathered my thoughts and met with Stewart, a very open and amenable guy. I was also keen to find out about and dispel some myths – I hadn't realised, for example, that there is a relatively even spread of pigeon fanciers across the UK, particularly associated with urban areas. I had always thought there was a greater concentration in the northern parts of England. It became very clear that pigeon fanciers are putting a lot of time, effort and money into their hobby, which for some is a livelihood.

The most significant development I heard about from Stewart was the racing pigeon unions joining together to form a coordinating voice, the Raptor Alliance. They have submitted an application to the government to change the law to recognise racing pigeons as livestock. This is contributing towards a review by the government on the wildlife laws that exist in the UK. If accepted, racing pigeons would be protected like pheasants and grouse. Individuals who have significant problems with their birds being killed or injured by raptors such as Sparrowhawks would be able to apply for a licence to mitigate the problem and remove the offending bird of prey by trapping it. I wasn't sure what would happen to the bird after this – would releasing it elsewhere work? I know Sparrowhawks don't do well in captivity. The preference is not to kill an offending bird, so further solutions as to where the bird goes after trapping would need to be considered. It was interesting to hear further that issues with hawks and Peregrines are very much regional – some places such as the Welsh Valleys are more vulnerable than areas in East Anglia where there are few, if any, problems. Racing pigeon routes pass over certain Peregrine territories but not others. Currently, members of the pigeon unions are encouraged to report genuine hawk attacks online. This will provide some evidence and facts, and not just hearsay or opinions. They are also gleaning information from racing rings found at Peregrine sites to see if they are from local or passage birds. The information provided to members does also state very clearly that it is illegal to shoot, poison or trap a raptor, and gives useful tips on reducing hawk attacks on their pigeons.

In theory, if the protection of pigeons is granted it doesn't mean that every racing pigeon fancier in the UK would be able to apply for a licence. Just as with applications by landowners to control Buzzards, licences would be controlled, and the application process would be tough and thorough. But it would no doubt be controversial for many. In 2013 it was revealed that Natural England had granted licences for four Buzzard nests and their eggs to be destroyed on a game shooting estate to protect pheasants. This was at a time when people were already in uproar that culling of Buzzards had even been considered by the government. There was understandably a huge

Figure 10.14 A pigeon loft used for racing pigeons in their first year of competing (Royal Pigeon Racing Association).

outcry from conservation organisations and individuals, with a feeling this was inappropriate, untimely and unacceptable. It sets a bad example during a period when the government and police need to be further prioritising raptor crime and stamping out the killing of birds such as Buzzards, rather than supporting their legal demise.

If the Raptor Alliance's submission goes through it will no doubt provide an opportunity for lots of discussion and debate as to how racing pigeons can best be protected from persistent hawks without them being killed but instead trapped and removed. But would it set a precedent? At a time when raptor persecution is at an all-time high, the message still needs to be positive one – that birds of prey deserve continued protection. Future changes and solutions with regard to racing pigeons will need further debate and discussions with all parties, including those involved with the conservation of raptors.

But how do you stop a Sparrowhawk or Peregrine eating a pigeon? The simple answer is you can't stop their natural instinct to catch prey such as pigeons. But there are simple steps which can help racing pigeon fanciers reduce the risk of their birds being predated. These range from flying birds from their loft at different times of the day to avoid hawks expecting the pigeons to be available at a set, regular time; changing the routes of flights to avoid known Peregrine sites; keeping pigeons in good condition and not overweight so they can escape a predator more easily; and training pigeons to return directly to their loft rather than spending time on the lawn, where they can be easily caught by a hawk (Fig. 10.14).

What does seem to be clear is that the main issue for pigeon fanciers is related more to Sparrowhawks and less to Peregrines. Young pigeons exercised locally near their loft are the main targets, and the hawks will also enter lofts. In areas where Sparrowhawks are regular visitors, pigeon attacks and deaths can mean a gradual decline and decimation of a flock. In urban areas, raptors often have small territories and live in much higher densities than they would in rural areas. As Sparrowhawks have increased in urban areas, so too have their densities and therefore their potential impact on vulnerable lofts.

Figure 10.15 A Peregrine with a freshly caught young pigeon (Hamish R. Smith).

Peregrines, meanwhile, cause more of an issue for racing pigeons when the birds are flying in races or exercising further away from their loft (Fig. 10.15). In response to a Peregrine coming in, flocks (especially those containing young birds) will scatter, with the knock-on effect of delaying their arrival back at the loft. Some may get predated and others may also be affected by weather and collisions.

Looking out for the safety and welfare of urban Peregrines

When Peregrines decide to come into towns, cities and other urban areas we feel compelled to look after them, to provide them with nesting sites, and to ensure their well-being. With many Peregrines still being persecuted in the UK, there is still an important task ahead to ensure the safety, survival and protection of the species.

We have a responsibility to develop a better understanding of how the Peregrine lives in the natural environment, to help dispel myths, and to ensure that illegal activities such as shooting and poisoning are dealt with (Fig. 10.16). The legal systems in the UK and beyond need to take raptor persecution more seriously, and those involved in the unnecessary slaughter need to face the consequences. For there to be long-term changes, and

Figure 10.18 Visiting a falconry show can be a fun and engaging way of showing children and their families falcons such as Peregrines (Hamish R. Smith).

2012, the RSPB put forward a list of recommendations relating to wildlife crime to the Law Commission in the UK, who have been reviewing species legislation in England and Wales. It is hoped that these will be taken on board and implemented to ensure wild birds are better protected, and that those who have committed wildlife crimes are properly prosecuted.

Peregrines and children

In urban areas Peregrines are on our doorstep. With little effort, Peregrines in many towns and cities can be spotted on the way to work, school or the shops. While Peregrine watches bring people closer to Peregrines, they are also a bird that others can happen upon, or see as part of their everyday life. As Stephen Moss outlined in the National Trust's report *Natural Childhood*, there is a perception that nature can't be accessed unless it is part of an organised activity, or that it requires equipment; that it has to be an interactive experience. But Peregrines don't need to be in a nature reserve or fenced off. Perhaps that is why they are so attractive – they don't require anything extra to be able to see them, and while the odd telescope or binoculars can be useful, their presence in the sky or on a building is enough to excite people.

At a time when children are glued to their smartphones, computer games and laptops, how are we able to engage them more with the outdoors, Peregrines, and other wildlife? The technology they are using may offer some solutions, such as accessing specific apps, websites and suggestions of things to do. Television provides the opportunity for children to see nature and discover more about it. For example, the BBC's series *Deadly 60* has been a huge hit with children in the UK and has provided a way in which they can find out more about the big, the small and the ugly, and what eats what!

Peregrines are of course included in this. It reminds me of *The Really Wild Show*, also featured on the BBC, which ran between 1986 and 2006. For me, this was the must-see show back in the 1990s, with presenters such as Chris Packham who still inspire me today.

But how about connecting children with the real outdoors, the fresh air, the touch and feel of nature, the sounds and smells of natural life? There are white papers written on the subject and initiatives and projects being delivered by non-governmental organisations (NGOs) and charities who are working hard to enable children to experience, learn about and engage with the natural environment, combining a whole variety of health, educational, community and environmental benefits.

If you would like to engage children with Peregrines, here are a few ideas of how they can see one or find out more about them in real life:

- Falconry centres offer a chance to see captive Peregrines up close, and often as part of a flying display. Birthday presents are often available for children (and adults) to fly birds from their hand (Fig. 10.18).
- Find out about local wild Peregrines that are well known and accessible – contact local bird clubs, search the location and the word 'peregrines' online, or check websites of larger organisations (in the UK, the RSPB and Hawk and Owl Trust). There may also be regular blogs following Peregrines at specific sites.
- If Peregrine watches are available to visit, just turn up and see the Peregrines. With the help of others, you may even get to see them through a telescope (Fig. 10.19). Under adult supervision, some children may be able to volunteer at a Peregrine watch (Fig. 10.20).
- Webcams – throughout the world there are various web cameras fixed on to Peregrine nests. And there are plenty in the UK, from Nottingham to Exeter.
- Museums – there is huge value in the specimens and handling objects that museums have both on display and during workshops or specific events. In the UK, many museums are free to visit and their exhibits often include natural history and mounted Peregrines.
- Boat trips – coastal and estuary trips to see dolphins and birds often result in spotting Peregrines, too. For example, in Bristol a trip down the Avon Gorge provides a great chance to see Peregrines both in the Avon Gorge itself and further down close to the Severn Estuary.
- Food webs – Peregrines are part of a bigger, wider web of animals and

plants and this provides a great example of how creatures interact and eat each other. Food webs are important concepts for children to grasp, and as top predators Peregrines can be used as a local, relevant case study both in school and at home.

Visiting schools

Showing people Peregrines at a Peregrine viewing point is ideal for members of the public. They can see a real Peregrine and get involved with the organisations showing them. However, if you have the opportunity to visit schools or deliver workshops to families away from Peregrine sites, it is worth keeping in mind a few things to ensure that both yourself and the children or young people get the most out of the experience.

Firstly, it is important to know exactly what you want your audience to learn from your visit – this will help inform you (and the teachers or event organisers) about what you want the children to get from the experience. Are they going to find out more about Peregrines? Are they going to find out more about how urban areas are good for wildlife? Will they discover how they can help wildlife where they live?

Then you need to think about the structure of a session and how long it will run for. Providing pictures, words, sounds, and objects for children and young people to handle will all help to ensure you reach most, if not all, of your audience. Feeling things is especially important as children learn through touching and exploring objects. As a general rule, a child's age relates to their attention span. So for a group of 5-year-olds you would want a session that is no more than 30–45 minutes long, with different activities or a switch in what you are doing every 5 minutes or so. However, with 11-year-olds you can provide activities and questions that are more challenging and last a little longer. You will provide a better experience if you are facilitating activities rather than simply talking in front of a group of children for half an hour or more – if the latter, the children will simply switch off.

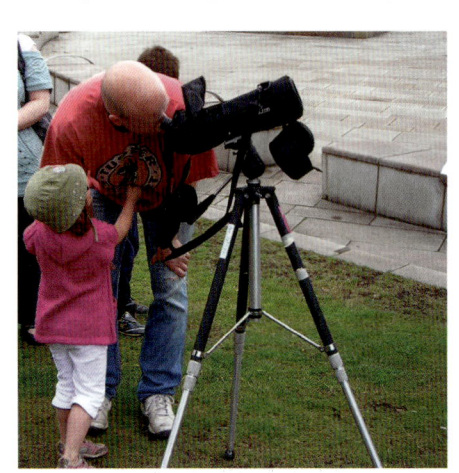

Figure 10.19 Enabling families to see real wild Peregrines is exciting and offers a 'wow' experience for everyone (Nick Brown).

Figure 10.20 Children love to see Peregrines up close and in action (Hamish R. Smith).

This will be a waste of both your time and their learning.

For an adult audience, illustrated presentations are acceptable but provide a passive, didactic transfer of information. However, this format is popular with bird clubs and societies, and is enjoyed by those who come along. It is difficult to offer something other than a talk at a regular club meeting, but providing a range of photos, sounds and words can all help to enhance the presentation and keep people engaged (and awake!).

Messages to people

When we communicate messages to the public about nature and wildlife it is very easy to state the facts and leave people depressed, with little thought about what they can do to make a difference. Conservation organisations have definitely got better at how they deliver messages through the media and at public events such as festivals and shows. However, there are still stories and facts which are portrayed as doom and gloom, with little hope for the viewer, reader or listener to feel empowered to do something to make a difference. This could include changing their behaviour or simply doing something such as a survey or making a home for wildlife. The report *Branding Biodiversity: The New Nature Message* produced by Futerra Sustainability Communications comes up with solutions to create change and sustainability in companies' and people's lives. Using physiological evidence to find out what drives people to conserve nature, the document outlines how to best convey messages depending on who you are speaking or delivering to. So, for example, if communicating messages about Peregrines, combining a mixture of celebrating and loving Peregrines with how people can watch them and do more to see or protect them will be far more effective than conveying messages of persecution and inaction, the doom and gloom elements. An organisation may therefore want to encourage the public to enjoy and love Peregrines, and then sign a petition or ask people to lobby against persecution. This is more empowering than telling people that Peregrines are shot each year and that we have fewer living on our grouse moors in the north. People are left feeling sad and with no creative ideas of how they can help.

There are also other ways of sending out positive messages. It is very easy to tell people to not do something or put up signs which simply say 'no dogs' or 'keep out'. Research in zoos in Australia has shown that using a different way of communicating simple messages works a treat. Like *Branding Biodiversity*, it is about understanding how and why people do things. How do we get everybody connected with nature and wanting to do something to help? The research, by Monash University, showed that zoo experiences were changing attitudes and arousing people's curiosity and interest in animals, but there was little evidence that this was influencing behaviour. So while visitors may experience a workshop or show about a particular animal, they weren't going away and changing their behaviour, despite hearing about the animal's decline. Changing behaviour instead requires empowering people and sending out positive messages about what you or the organisation has done or is doing. For example, 'Protect peregrines by lobbying the government to stop illegal persecution' may create very little behavioural change compared to telling people, 'We've already sent 25,000 names to the government and donated £100,000 for the protection of upland moors. Please sign this petition…'. When people know you have already invested something, they are more likely to follow suit.

Chapter Eleven

Where Next?

When Derek Ratcliffe wrote his book *The Peregrine* in 1993 there was only a short mention about urban Peregrines, nothing that would warrant the writing of a whole book about them. The following 20 years saw a sudden rise in the population of urban Peregrines in towns and cities across Europe and other parts of the world.

So what will the next 20 years bring for Peregrines? Are there things I have only mentioned in this book which in 2035 will be far more commonplace? Will Peregrines be even more common, or plateau? Only time will tell, and much is dependent on how well we protect and conserve them and their

Figure 11.1 In many towns and cities you are more likely to see a Peregrine than a Sparrowhawk or Kestrel (Sam Hobson).

environment. However, one thing is for sure – if Peregrines continue to exploit the urban environment both in the UK and around the world without being persecuted, they will be doing better than they have done for thousands of years (Fig. 11.1).

While persecution still persists, their Schedule 1 protection in the UK is paramount to ensuring that Peregrines remain undisturbed, better understood and alive. Without this, complacency may see pockets of persecution and disturbances happen more regularly. Chemicals in the countryside also remain a threat to the Peregrine and need to be monitored to avoid the calamity we witnessed over 50 years ago with DDT.

As funding is prioritised for Peregrines, particularly in places where they are still rare and vulnerable, we can build up an accurate picture of just how this falcon lives. Satellite tracking is helping to find out which countries and habitats they use on migration, what borders they pass through, how long it takes them to complete their journeys, and how known threats such as persecution or changes in certain habitats may affect them. And as mobile satellite technology gets smaller and smaller, the ability to tag Peregrines with reliable satellite data loggers at an affordable price is becoming more realistic. Even Peregrines in remote places can be followed and give us a better understanding of their lives and survival.

Ringing of Peregrines is becoming more widespread, helping us to find out about the interwoven lives of different individual Peregrines. As Peregrines become more accessible in urban areas, more individuals can be ringed. Additionally, ringing projects specifically studying Peregrines are expanding around the world. As Peregrines become increasingly seen and observed, there is a real opportunity to learn more about who they mate with, how many partners they have, and other behaviours which may change in response to a growing, dynamic population (Fig. 11.2).

Web cameras continue to be popular, and as technology develops further and becomes cheaper, their high definition and clarity will provide a better experience for the viewer, helping the urban Peregrine to continue to entertain and enthuse people (Fig. 11.3).

Figure 11.2 Ringing of Peregrine chicks will continue to tell us more about their movements, breeding partners and survival (Ian Sparrowhawk).

Figure 11.3 Recording Peregrine behaviour using web cameras and streaming it across the Internet or broadcasting it on television is becoming more and more popular.

We have a good understanding of what Peregrines eat, but finding out more about where Peregrines range, where they are feeding, and important foraging sites would better improve our understanding of what makes a Peregrine tick. Are declines in some parts of the UK, particularly in Scotland, related to foraging success and availability of prey, for example? Or are other factors such as climate change or persecution to blame? Or is it a combination of them all?

Wildlife hospitals will encounter more young Peregrines which are found grounded, either healthy or injured. And adults will also find themselves in trouble from time to time. Conflicts with people and Peregrine nest sites will increase or continue in towns and cities, especially where buildings (and therefore nest sites) may be pulled down; building work is delayed; residents in blocks of flats are inconvenienced; gutters get full or blocked by

Figure 11.4 A Peregrine brings food to a nest overlooking a busy road (Pete Blanchard).

Figure 11.5 A juvenile Peregrine newly fledged from its nest in the Avon Gorge in Bristol (Sam Hobson).

prey items; and misunderstandings about Peregrines and their prey occur. For example, in some locations Peregrines get blamed for the piles of mess that Feral Pigeons make. However, with good local liaisons with researchers, Peregrine enthusiasts, local bird clubs, conservation organisations and the media, many of these conflicts can be mitigated or completely removed with time, patience, diplomacy, and a view of the bigger picture and both sides of the argument.

Peregrines will hopefully continue to thrive in our towns, cities and industrial areas, as well as their more traditional haunts in quarries, rocky crags and moorland (Fig. 11.4 and 11.5). They are a great bird to get people interested in nature, and often people know of a Peregrine and its speed even if they have little knowledge of or interest in other wildlife (Fig. 11.6 and 11.7).

If you would like to get more involved with Peregrines, or would like to watch them from your own living room, here are some organisations and links to resources which may be useful:

Web cameras
- Derby – derbyperegrines.blogspot.com
- Norwich – upp.hawkandowl.org/

Figure 11.6 An immature Peregrine – ideally suited for a predatory lifestyle.

- Brighton – search online for 'Sussex Heights' and 'Peregrines'
- Nottingham – search online for 'Nottingham University' and 'Peregrines'
- Worcester – worcester.gov.uk/peregrine

Organisations
- Hawk and Owl Trust – hawkandowl.org
- Royal Society for the Protection of Birds – rspb.org.uk
- And for children, RSPB Wildlife Explorers – rspb.org.uk/youth/join_in/wex.aspx
- European Peregrine Falcon Working Group – falcoperegrinus.net
- London Peregrine Partnership – london-peregrine-partnership.org.uk (includes links to other peregrine groups and blogs)
- Raptor Politics – raptorpolitics.org.uk
- Hawk Conservancy Trust – hawk-conservancy.org
- The International Centre for Birds of Prey – icbp.org
- Scottish Raptor Study Group – scottishraptorstudygroup.org
- Welsh Raptor Study Group – birdsinwales.org.uk/rare/wrbbrsg.htm

- Northern Ireland Raptor Study Group and Raptor Monitor – raptormonitor.com
- Irish Raptor Study Group and The Golden Eagle Trust – goldeneagle.ie
- British Trust for Ornithology – bto.org
- The Peregrine Fund – peregrinefund.org
- Raptor Research Foundation – raptorresearchfoundation.org (includes links to the Journal of Raptor Research)
- Global Raptor Information Network – globalraptors.org
- BBC Nature: Peregrine falcon – bbc.co.uk/nature/life/Peregrine_Falcon
- ARKive: Peregrine falcon – arkive.org/peregrine-falcon/falco-peregrinus
- Royal Pigeon Racing Association – rpra.org
- Game & Wildlife Conservation Trust – gwct.org.uk
- National Trust: 50 things to do before you're 11 ¾ – 50things.org.uk
- STEMnet – stemnet.org.uk (creates opportunities to inspire young people in Science, Technology, Engineering and Mathematics)

Figure 11.7 A female Peregrine heading down to a lower position on a church (Hamish R. Smith).

Further reading

Annual Reports of the Peregrine Protection Study Group, North Rhine-Westphalia (in German) – nrw.nabu.de/tiereundpflanzen/wanderfalke/jahresbericht

Avery, M. 2012. *Fighting for Birds: 25 Years in Nature Conservation*. Exeter: Pelagic Publishing.

Baker, J.A. 2011.*The Peregrine*. Harper Collins, London.

Baker, K. 1993. *Identification Guide to European Non-passerines*. BTO Guides 24. BTO, Thetford, Norfolk.

Balmer, D., Gillings, S., Caffrey, B., Swann, S., Downie, I. & Fuller, R. 2013. *Bird Atlas 2007–11: The Breeding and Wintering Birds of Britain and Ireland*. British Trust for Ornithology, Thetford, Norfolk.

Birkhead, T. 2013. *Bird Sense: What It's Like to be a Bird*. Bloomsbury Publishing Plc, London.

Branding Biodiversity: The New Nature Message by Futerra Sustainability Communications available at http://www.futerra.co.uk/downloads/Branding_Biodiversity.pdf

Brown, R., Ferguson, J., Lawrence, M. & Lees, D. 2003. *Tracks and Signs of the Birds of Britain and Europe*, 2nd Edition. Christopher Helm, London.

Cieślak, M. & Dul, B. 2006. *Feathers: Identification for Bird Conservation*. Natura Publishing House, Warsaw, Poland.

Dixon, N. & Shawyer, C. *Peregrine Falcons: Provision of Artificial Nest Sites on Built Structures* by London Biodiversity Partnership available at http://www.lbp.org.uk/downloads/Publications/Management/peregrine_nest-box_advice.pdf

Drewitt, E.J.A. & Dixon, N. 2008. Diet and prey selection of urban-dwelling Peregrine Falcons in southwest England. *British Birds* 101: 58–67.

Easlea, B. 2008. *Birdwatching at the Seaside: Living with Peregrines and Other Birds in a Sussex Coastal City*. Pen Press Publishers, Brighton.

FALCO, the newsletter of the Middle East Falcon Research Group – mefrg.org

Fanshawe, J. (ed) 2011. *The Peregrine: The Hill of Summer & Diaries: The Complete Words of J.A. Baker*. HarperCollins, London.

Frank, S. 1994. *City Peregrines: A Ten-year Saga of New York City Falcons*. Hancock House Publishers Ltd, Canada.

Jenni, L. & Winkler, R. 1994. *Moult and Ageing of European Passerines*. Academic Press Limited, London.

Hardey, J., Crick, H., Wernham, C., Riley, H., Etheridge, B. & Thompson, D. 2009. *Raptors: A Field Guide for Surveys and Monitoring*, 2nd Edition. The Stationery Office Ltd, Edinburgh.

Macdonald, H. 2006. *Falcon*. Reaktion Books Ltd, London.

Moss, S. 2012. *Natural Childhood*. National Trust, UK.

Olsen, J., Fuentes, E., Bird, D.M, Rose, A.B. & Judge, D. 2008. Dietary shifts based upon prey availability in Peregrine Falcons and Australian Hobbies breeding near Canberra, Australia. *J. Raptor Res*. 42 (2): 125–137.

Oro, D. & Tella, J.L. 1995. A comparison of two methods for studying the diet of the Peregrine Falcon. *J. Raptor Res*. 29: 207–210.

Ratcliffe, D.A. 1993. *The Peregrine Falcon*, 2nd Edition. Poyser, London.

Scott, S.D., & McFarland, C. 2010. *Bird Feathers: A Guide to North American Species*. Stackpole Books, Mechanicsburg, Pennsylvania, US.

Stirling-Aird, P. 2012. *Peregrine Falcon*. New Holland, London.

Sielicki, J. & Mizera, T. (eds) 2009. *Peregrine Falcon Populations: Status and Perspectives in the 21st Century*. TURUL / Poznań University of Life Sciences Press, Warsaw / Poznań.

Takada, M. & Kanouchi, T. 2004. *The Feathers of Japanese Birds in Full Scale*. Bun-Ichi Co., Ltd., Japan.

Tucker, V., Tucker, A.E., Akers, K. & Enderson, H.J. 2000. Curved flight paths and sideways vision in peregrine falcons (*Falco peregrinus*). *J. Exp. Bio*. 203 (24): 3755–3763.

Wernham, C.V., Toms, M.P., Marchant, J.H., Clark, J.A., Siriwardena, G.M. & Baillie, S.R. (eds) 2002. *The Migration Atlas: Movements of the Birds of Britain and Ireland*. Poyser, London.

For children

Arnold, C. 1985. *Saving the Peregrine Falcon*. First Avenue Editions, Minneapolis. (Available in the UK through nhbs.com)

Bradley, M. 2013. *Top Gun of the Sky*. Ceratopia Books, New Forest.

Dunphy, M. 2000. *The Peregrine's Journey: A Story of Migration*. The Millbrook Press, Brookfield, Connecticut. (Available in the UK through nhbs.com)

Lunis, N. 2010. *Peregrine Falcon: Dive, Dive, Dive*. Bearport Publishing, New York.

Taylor, M. 2010. *RSPB British Birds of Prey*. Christopher Helm Publishers Ltd, London.

Unwin, M. 2005. *Peregrine Falcon (Animals Under Threat)*. Heinemann, Portsmouth.

Wechsler, D. 2001. *Peregrine Falcons*. The Rosen Publishing Group, New York.

Acknowledgements

Huge thanks to my family and close friends, including Mike Dilger, Louisa and Daniel Aldridge, and especially my fiancée Liz Shaw. A huge thank you to all the photographers who have kindly allowed me to use their photos to bring the book alive with images.

And special thanks to everyone who has worked with me, provided support and advice, and without whom this book would not have been possible: Nigel Massen, Tom Newman, Thea Watson, Nick Dixon, Adrian George, Hamish Smith, Sam Hobson, Andrew Dixon, Graham Roberts, Denis Corley, Louise Hazleton, Stephen Moss, Peter Wegner, Janusz Sielicki, Mike Rogers, Colin Morris, Andy Grant, Luke Sutton, Nick Brown, Nick Moyes, David Morrison, Bob Carling, Stuart Harrington, Rainer Altenkamp, Dave Grubb, Stewart Waldrop, Alice Tribe, Mark Thomas, Chris Packham, John Tully, Richard Bland, Peter Welsh, Mandy Leivers, Bill Morris, Robert DiCandido, Dr Rob Thomas, Lloyd and Rose Buck, British Mountaineering Council (especially Colin Knowles, Rob Doman, Daniel Donovan and Simon Fletcher), Bristol Ornithological Club, Bristol Naturalists' Society, Peter Rock, Mike Bailey, Robin Prytherch, Chris Jones, Adele Powell, Nik Ward, Peter Howlett, Allison Yellowley, Rupert Griffiths, Paul Oaten, Shawn Clements, Stephen Woollard, Chris Sperring, Terry Pickford, Dave Gittens, Kim Kirkbridge, Ashley Smith, Caro Catoni, Tuomo Ollila, Sergio Seipke, Craig Bell, Marc Ruddock, John Black, Jim Wells, Victor Hurley, Jerry Olsen, Oscar Beingolea, Judith Smith, Terry Rowe, Mike Hunter, Edmund Flach, Denzil Large, Tracey Rich, Tim Aldred, RSPB, RSPCA, Hawk Conservancy Trust, British Trust for Ornithology, Hawk and Owl Trust, North Cotswold Ornithological Society, St John's Church (Bath), British Birds, BBC Wildlife Magazine, Royal Pigeon Racing Association.

Photographers

Sam Hobson

Sam Hobson is a wildlife photographer based in Bristol. He has watched Peregrines all over the UK, but his favourite place to photograph them is in cities, particularly in London and Bristol. 'The best thing about urban wildlife is that it can be habituated and comfortable with the presence of man. In the city there are opportunities to get close enough to study and photograph species that would be next to impossible in the countryside, and that includes Peregrines. I like to show the context of the animal as much as possible, and an urban background full of man-made structures not only helps to tell the story of our relationship with these raptors, but also makes for interesting and often artistic compositions.' Sam is continuing to work with urban Peregrines and is also now working on a project on urban Goshawks in European cities.
samhobson.co.uk

Hamish Smith

For Hamish Smith wildlife, travel and photography have always featured highly on his list of interests and hobbies. In his early years growing up in Scotland he combined all three with a love, if not matched in ability, for fly fishing, travelling to the remoter areas of the north. Throughout his 40+ years as a professional engineer, while cricket, reading, music and theatre fitted more comfortably with family life and the demands of the day job, photography remained a constant. Just ask his family!

These days, while he retains an enduring interest in the specialisms of his past life, his passion for his personal hobby trinity of wildlife, travel and photography has been refreshed. He is an active member of the Hawk Conservancy Trust, the Hawk and Owl Trust, for which he is the Project Coordinator for the Bath Urban Peregrine Project, and the British Trust for Ornithology, under whose umbrella he is being trained by the author of this book to ring Peregrines.

Over the years he has developed a particular interest in Peregrines, and the camera system installed on the Bath nest box is providing Hamish with a fascinating insight into the day-to-day life of a mature breeding pair of these stunning creatures.

In keeping with his hairline, Hamish's gear has moved on too. Gone are the days of Zorki and Zenit. These days his tools of choice hark from the Canon stable, and a 1D MkIV and 5D MkII can normally be found attached to a Canon 500mm f4L MkII and a 70-200 f2.8L MkII IS, backed up by a range of shorter L-series lenses. Having said that, his Canon 1nHS and Bronica film cameras are never far from hand.

hamishsmithphotography.co.uk

Dave Pearce

Dave Pearce spends most of his birding activities in the under-watched area of the countryside of the North Cotswolds, with farmland birds being a particular interest. He carries out BTO surveys and liaises with local farmers in the area. The Severn Estuary and the Forest of Dean are also places he visits for other special Gloucestershire birds. After setting up a Peregrine box on a local church with a CCTV feed to his home, Dave has had over the last few years a privileged view of the life of a pair of urban Peregrines. Although he would not refer to himself as a photographer, it has also allowed him to take close-up photos without the birds being aware of him.

ncosbirds.org.uk/peregrines.html

Paul Williams

Paul Williams is a wildlife photographer, and TV producer/director at the BBC Natural History Unit. Many of his images can be seen at:

ironammonitephotography.com

Allan Chard

Allan Chard is a keen amateur wildlife photographer around Bristol, mainly capturing stunning bird images, but also photographing any wildlife that lingers long enough for him to focus on.

Slawomir Sielicki

Slawomir Sielicki is a forester by education, a falconer of 30 years, and for the past 18 years has professionally been involved in nature conservation, mainly restoring the Peregrine population in Poland. He is co-founder and President of the Society for Wild Animals, 'Falcon'.
peregrinus.pl

Peter Wegner and Thorsten Thomas

Peter Wegner is the founder, leader and speaker of the Peregrine protection group of North Rhine-Westphalia, Germany, set up 25 years ago. Peter works alongside Thorsten Thomas to monitor Peregrines in this region. In 2013 the group monitored 189 pairs in their federal state, which produced 339 chicks, of which 233 were ringed. For the group's 25th anniversary a special book was produced about their work, successes, and changes in the Peregrine populations during that period.

Jen Bright

Jen Bright is a research scientist investigating the evolution and biomechanics of bird skulls.
bristol.ac.uk/earthsciences/people/jen-a-bright/index.html

Adam Rogers

Adam Rogers is an urban ecologist with a passion for revealing the hidden lives of some of our most familiar urban wildlife.
adamrogersonline.com

Jason Kernohan

Jason Kernohan is a local patch birder, amateur naturalist and wildlife blogger based in Worcestershire, Central England.
shenstonebirder.blogspot.co.uk

Andy Thompson

Andy Thompson is an amateur wildlife photographer based in Norfolk. He is involved with the Hawk and Owl Trust and the Norwich Peregrine Project, and has spent many hours watching, recording and photographing the Peregrines at Norwich.

Pete Blanchard

Pete Blanchard has been taking digital wildlife photos for about ten years and has been interested in wildlife since watching Peter Scott and David Attenborough on television back in the 1950s. He takes photos mainly in south-west England and birds of prey in flight are his favourite subject.
flickr.com/photos/flyingfast

Nathalie Mahieu

Nathalie Mahieu is a Peregrine enthusiast and has been monitoring the Peregrines on Charing Cross Hospital in West London since the falcon arrived in autumn 2007.
facebook.com/FaBPeregrines

David Gittens

David Gittens has a technical background in engineering and science, a lifetime interest in wildlife and nature and a keen interest in birds of prey. David looks after the peregrine nest CCTV system at Norwich Cathedral and downloads images and video clips so that, under guidance from the Hawk and Owl Trust, they can be posted on YouTube and Facebook.
upp.hawkandowl.org

Nick Brown

Nick Brown of Derbyshire Wildlife Trust (holding a grounded juvenile) was alerted to the presence of Peregrines on Derby Cathedral in 2004/5 and soon saw the potential to put up a nest platform which was achieved in 2006. The same pair has raised over 30 chicks since then.
derbyperegrines.blogspot.co.uk

Nick Moyes

Nick Moyes, a mountaineer and IT wizard, set up the cameras in Derby city centre in 2007, and by 2014 over 2.6 million hits had been registered. Nick was formerly the curator at Derby Museum & Art Gallery and now works as an ecological consultant.
@nickmoyes

Elizabeth Woodward

Elizabeth Woodward works for Derbyshire Wildlife Trust and has a keen interest in Peregrines, especially urban ones.
derbyshirewildlifetrust.org.uk

Ian Sparrowhawk

Ian Sparrowhawk is an amateur photographer who inherited a love of wildlife from his Dad while growing up in rural Oxfordshire. Now living in Bath it is a real treat to see peregrines daily, regularly skimming over the building tops of the city centre.

Chris Jones

Chris Jones is a Bristol-born photographer, and happy there isn't a cure for his Peregrine obsession. He spends time photographing the Avon Gorge Peregrines and produces stunning images.

Gary Thoburn

Gary Thoburn has been birding since the age of four when his dad used to go birding around his local wood and very small reservoir in the Lawrence Weston area of Bristol. Photographing birds from an early age, Gary has travelled all around the UK and many parts of the world photographing many species, and has an incredible portfolio on his website: **garytsphotos.zenfolio.com**

Ronald van Dyke

Ronald van Dyke studies Peregrines in the Netherlands and says, 'Being face to face with a male Peregrine is a lifetime experience.'

Matt Allen

Matt Allen is a 20-year-old aspiring professional wildlife photographer, filmmaker and wildlife guide. He is just finishing a Biology degree at the University of Bristol, and during his summers he has ventured to the great Alaskan and Canadian wildernesses to professionally guide photographers and wildlife enthusiasts to see the charismatic Brown Bear.
mattallenwildlife.com

Christine Raaschou-Nielsen

Christine Raaschou-Nielsen lives in rural southern Denmark and loves photographing wildlife and landscapes across the UK and Scandinavia, spending many hours in dome-hides as well as roaming forests, moors and beaches. She works with a Canon EOS D5 MKII and MKII with various lenses.
northshots.net

Dave Grubb

Dave Grubb runs Electrodesigns CCTV Limited and photographs wildlife as a hobby. He helps oversee cameras on a church in Worcester city centre that is used by different Peregrines including a regular wintering female known as Bobbin.
worcester.gov.uk/peregrine
facebook.com/worcesterperegrines

Mike Wallen

Mike Wallen has lived in Bucks all his life and has been birding the county seriously for 25 years. He currently edits the Annual Bird Report (and has done so for more years than he cares to remember). Every spring he visits the wonderful Skomer Island and counts the gulls there. His special interests are visible migration and of course bird number one – the Peregrine. He is heavily involved in the Aylesbury Peregrine Project and keeps a note of the birds' diet, which has included a type of pipistrelle bat.
biodiversity.aylesburyvaledc.gov.uk

Liz Shaw

Liz Shaw is a zoologist and freelance proofreader, editor and writer with a passion for the natural world. She can often be found roaming the countryside with her camera, photographing and blogging about the wildlife she finds. Liz also helps to run the Garden Bioblitz, a national event encouraging people to get outside and discover the wildlife living on their doorstep. She is a keen artist in her spare time.
ramblingsofazoologist.com
lizshaweditorial.com

David Lindo

David Lindo is The Urban Birder – a writer, broadcaster, speaker and bird guide. His whole vibe is about getting urbanites to realise that there is a wealth of wildlife under their noses in the world's cities.
theurbanbirder.com

Index

Page numbers in *italics* refer to figures.

adaptability 8–9
adaptations *10*, 10, 15, 22
aerial views *172*, 172
African Grey Parrot (*Psittacus erithacus*) 90
age 32
American Woodcock (*Scolopax minor*) 98
The Annals of Scottish Natural History 148
Arctic Tern (*Sterna paradisaea*) 87
Argentina 36, 97
Atlas of Birds of Aragon 108
auks 88
Australia 7, 57, 68, 84, 98, 186
Avery, Mark 174
Avocet (*Recurvirostra avosetta*) 87
Avon Gorge, Bristol, England 57, 70, 77, 121, *134*, 164, *190*
Aylesbury, England 50–51, *118*

Baillon's Crake (*Porzana pusilla*) 97
Bar-headed Goose (*Anser indicus*) 22
Barn Owl (*Tyto alba*) 89
Barn Swallow (*Hirundo rustica*) 72, 92
Barnacle Goose (*Branta leucopsis*) 140
Bath, England
    cooperative breeding 67, 67–68, *74*
    diet 98, 99
    grounding 152, 154
    gulls and raptors 102
    hunting 78
    media involvement *117*, *134*
    perching spots *41*, 43
bats 92, 92–93
bazas 102
BBC 68, *117*, *133*, 133–35, *134*, 182–83
bees 151, *151*
bill *6*, 6, 76, 80, *80*
biometrics *129*, 131–32
biophilia 165
Bird Atlas 2007–11 3
Bird Crime (RSPB) 181
Bird Feathers (Scott and McFarland) 108
BirdLife International 161
bitterns 91
Black Baza (*Aviceda leuphotes*) 102
Black Eagle (*Ictinaetus malayensis*) 102
Black-headed Gull (*Chroicocephalus ridibundus*) 88, 88, 93, 143, *143*
Black-necked Grebe (*Podiceps nigricollis*) 91, 94, *101*, 101

Black-tailed Godwit (*Limosa limosa*) 87
Black Tern (*Chlidonias niger*) 87
Blackbird (*Turdus merula*) 78, 79–80, *84*, 85, *157*, 157
Blackcap (*Sylvia atricapilla*) 85
Blue Tit (*Cyanistes caeruleus*) 17–18, 72, 85
boat trips 183
Bohemian Waxwing (*Bombycilla garrulus*) 86
Brambling (*Fringilla montifringilla*) 86, 95
Branding Biodiversity (Futerra Sustainability Communications) 119–120, 185
Bright, Jen 13
Bristol, England
    cooperative breeding 69, 70
    foxes *167*, 167
    landscapes 33
    nest boxes *116*, 116
    pigeons 172
    prey 75–76, 93
    ringed falcons 51
    roofs 37
    urban sites, use of *42*, 44
    wind turbines 159–160
    see also Avon Gorge, Bristol, England
British Mountaineering Council 124, *134*
British Trust for Ornithology (BTO) 3, 54, 57, 128, 131, 135
brominated flame retardants (BFRs) 150
brow 11
Brown Falcon (*Falco berigora*) 7
Buck, Lloyd 20
Budgerigar (*Melopsittacus undulatus*) 90
Bulgaria 39, 149, 161
Buzzard (*Buteo buteo*) 21, 22, 90, 102, 149, 155, 177–78

caches, prey 44, 52–53, *53*, 79, *100*, 100–101
caged / escaped birds 90, *90*
camouflage *41*, 42
Canada 27–28, 36, 68, 72, 118, 139
captive breeding programmes 146
cardio-vascular system 22
carotenoids 30–31
Carrion Crow (*Corvus corone*) 89
catching prey *6*, *65*, 65, 76, 80, 142
Catoni, Carlo 68
cere 11, *11*, 31, 32
Chaffinch (*Fringilla coelebs*) 86
chats 86
Cheltenham, England *31*, 39, *115*, 116, *118*
Chichester, England 46, 49

chicks 58–66, *59*, *60*, *61*, *62*, *63*, 147
 see also juveniles; ringing / tagging
Chiffchaff (*Phylloscopus collybita*) 85
children 122, *182*, 182–85, *184*, *185*
Cockatiel (*Nymphicus hollandicus*) 90
cockatoos 84
Collared Dove (*Streptopelia decaocto*) 82–83
colour perception 16–18
Common Cuckoo (*Cuculus canorus*) 126, 127
Common Gull (*Larus canus*) 78, 143, *143*
Common Sandpiper (*Actitis hypoleucos*) 87
Common Snipe (*Gallinago gallinago*) 94, 95
Common Swift (*Apus apus*) 19, *65*, 80, 92, *92*
Common Tern (*Sterna hirundo*) 34, *34*, 87
communicating messages 185–86
Computer Tomography (CT) *12*, 13
conflict resolution 174–79
cooperative breeding 66–72, *67*, *69*
Coot (*Fulica atra*) 94, 98
copulation 53, 55–56, *56*, *71*, 72, 125
Corncrake (*Crex crex*) *94*, 94, 95, *158*, 158
corvids 88, *89*, 89, *102*, 102–3, 143, *143*
crakes 91, *94*, 94–95, 97, *158*, 158
crimes against Peregrines 46–47, 147, 180–82
cuckoos 94, 97, 126, 127
Czech Republic 3, 39, 161

'A Date with Nature' watches 166
DDT (dichlorodiphenyltrichloroethane) 149–150
Deadly 60 (BBC) 182
death, causes of 146–47, 148–151
Dekker, Dick 77
Dell'Omo, Giacomo 68
Derby, England 57, 87, 95, 97, *117*, 118, *121*, 162, 166
diet studies 104
 collecting / sorting remains 110–14, *111*, *112*, *113*
 identification resources 107–8, 112
 identifying kills 108–10, *109*
 pellets *104*, 105, *105*
 prey remains 105–7, *106*, *107*, *110*
Dilger, Mike *133*, 133
Dipper (*Cinclus cinclus*) 90
distribution 3, *4*, 8–9
disturbances 161
Dixon, Andrew 137, 176
Dixon, Nick 3, 93, 94, 102, 116
Dorchester, England 44
doves 82–83, 95, 127
ducks *17*, 17, 90–91, *91*, 95, 105, *106*, 109, *110*, 142
Dunnock (*Prunella modularis*) 85, 95

Eagle Owl (*Bubo bubo*) 40, *40*, 76
eagles 102, 142, 164, 165
Eastern Rosella (*Platycercus eximius*) 84
eggs *54*, 54–55, 56–59, *57*, *58*, *59*, 150
egrets 91
Eleonora's Falcon (*Falco eleonorae*) 5, 9
England *see specific sites*
Eurasian Nightjar (*Caprimulgus europaeus*) 86, 127

evolution 5
Exeter, England 93, 102
extra-pair copulation 72, 125
eyes 11, *11*, *13*, 13–14, 18, 141
 see also sight

facial markings 11
falconry centres 183
falcons 5, 7, 8, 9, 24, 39, 160, 165
 see also Kestrel (*Falco tinnunculus*)
Feather Atlas (US Fish and Wildlife Service) 108
feathers *114*, 114–15
 see also moulting; preening
Feathers (Cieślak and Dul) 108
The Feathers of Japanese Birds in Full Scale (Takada and Kanouchi) 108
feeding 80, *81*
Feral Pigeon (*Columba livia*) *see* pigeons
Fieldfare (*Turdus pilaris*) 75, 85, 86, *86*
Fighting for Birds (Avery) 174
finches 17–18, 30, 86, 95
Finland 3, 18
flickers 94, 97
flight 15–16, 19–22, *20*, 42, *79*, 80
food webs 183–84
foods *see* prey
fovea 14
foxes 101, 166–67, *167*
future 187–190

g-forces 21–22
Gadwall (*Anas strepera*) 91
Galah (*Cacatua roseicapilla*) 84
gallinules 97
Gang-gang Cockatoo (*Callocephalon fimbriatum*) 84
Gannet (*Morus bassanus*) 88
geese 22, 101, 140
gene pool 5, 72
genome 5
George, Ade 125
Germany
 artificial nests 116, 117
 breeding studies 68, 69, 70, 72, 73
 gene pool 5
 population 3
 ringing studies 51, 136
 urban sites, use of 39–40
Gittens, David 61
godwits 87
Goldcrest (*Regulus regulus*) 85
Golden Eagle (*Aquila chrysaetos*) 142, 165
Golden Oriole (*Oriolus oriolus*) 86
Golden Plover (*Pluvialis apricaria*) 75, 87, *87*, 100
Goldfinch (*Carduelis carduelis*) 86
goshawks 11, 76, 102, 142, 164, 165
Great Bittern (*Botaurus stellaris*) 91
Great Grey Shrike (*Lanius excubitor*) 86, *113*
Great Tit (*Parus major*) 72, 85
Greater Painted-snipe (*Rostratula benghalensis*) 98
grebes 91, *91*, 94, 97, 98, *101*, 101, 142
Greenfinch (*Carduelis chloris*) 30, 86

Index

Grey Heron (*Ardea cinerea*) 91
grounding 45, 46, 62, 63, *152*, 152–53, 189
gulls *88*, 88
   attacks on 8–9, *78*, 102
   as predators 38, 102, 144
   as prey *88*, 88, 93, 143, *143*
   urban 36, *37*, 37–38, 95
Gyr Falcon (*Falco rusticolus*) 5, 8

harriers 31, 174
Harris, Stephen 167
Hawfinch (*Coccothraustes coccothraustes*) 86
Hawk and Owl Trust 46, 102, 166
Hawk Conservancy Trust 153
hawks 1, 11, 76, 142, 165
   see also goshawks; Sparrowhawk (*Accipiter nisus*)
head movements 14, 14–16, *16*
health indicators *30*, 30–31
heart rate 22
Hen Harrier (*Circus cyaneus*) 174
herons 91
Herring Gull (*Larus argentatus*) 36, 37, 38, *38*, 88, 143, *143*, 144
hirundines 92
Hobby (*Falco subbuteo*) 5, 9, 165
honeymoon period 154–55
House Sparrow (*Passer domesticus*) 84, 85, 167
hovering 101, 142
Hungary 39, 49, 138
hunting 76–80, *77*, *78*, 94–95, *96*, *97*, 97–101
Hurley, Victor 68
hybrids 8–9, 160–61

Identification Guide to European Non-passerines (BTO) 131
inbreeding 73–74, *74*
indentifying individuals *50*, 50–51
injured birds 46–47, 152–53, *153*, 189
Internet *117*, 117–120, *118*, *119*, 163, *189*
intruders 46, 53
Ireland 4, 93–94
Italy 38, 68–69, 78, 86, 92

Jack Snipe (*Lymnocryptes minimus*) 94–95
Jackdaw (*Corvus monedula*) 89
Japan 68, 70, 108
Jay (*Garrulus glandarius*) *89*, 89
Jenkins, Andrew 7
juveniles *191*
   accidents and injuries 152–53, *153*
   adventurousness 154
   begging *63*, 63
   catching prey *65*, 65
   cooperative breeding 67, 67–70, *69*
   dispersal 66
   feeding *61*, *64*, *65*, 65, *76*
   fledging 62–63, *63*
   grounding 45, 46, 62, 63, *152*, 152–53
   gull attacks 38, *38*
   mortality rate 32, 79
   moulting 25, *25*, *26*
   plumage 25, *25*, *26*, *31*, 42

skin colour 11, 28, *31*, 32
   see also chicks

Kestrel (*Falco tinnunculus*) 5, *21*
   decline 145
   hovering *142*, 142
   mate selection 24, 30–31
   pellets 109–10
   as prey 89
   sight 15, 18
   size 22
killing of prey 6–7, *80*, 80, 142
Kingfisher (*Alcedo atthis*) 90, *91*
Kirkbride, Kim 153
kites 149, *154*, 154–55
Kittiwake (*Rissa tridactyla*) 87
Klemann, Michel 108
Knot (*Calidris canuta*) 87

landowners 123–24, 173, 174
Lanner Falcon (*Falco biarmicus*) 8, 160
Lapwing (*Vanellus vanellus*) 75, *86*, 87
larks 95
laws 147–48, 180, 181–82
Leach's Storm Petrel (*Oceanodroma leucorhoa*) 88
Legal Eagle (RSPB) 181
leg colour 28, *29*, 30–32
Lesser Black-backed Gull (*Larus fuscus*) 37, *37*, 38, 88, *143*, 143, 144
Lesser Kestrel (*Falco naumanni*) 5
Lesser Redpoll (*Carduelis cabaret*) 86
life-span 32
lighting, urban 95, 99–100
Lind, Olle 18
Lindo, David 165
Little Auk (*Alle alle*) 88
Little Bittern (*Ixobrychus minutus*) 91
Little Egret (*Egretta garzetta*) 91
Little Grebe (*Tachybaptus ruficollis*) 91, *91*, 94, 101, *101*
Little Owl (*Athene noctua*) 89, *89*
Little Tern (*Sterna albifrons*) 87
London, England 37, 155–57, *156*, 166
London Peregrine Partnership 156, 157
Lothian and Borders Raptor Study Group 32, 132, 136
Lund University 18

magnetoreception 19
Magpie (*Pica pica*) 89
Malaysia 102
Mallard (*Anas platyrhynchos*) 17, *17*, 91
Manx Shearwater (*Puffinus puffinus*) 88
markings 11, 42
martins 92
mate selection 24, 30–31
mating 53, 55–56, *56*, 71, 72–74
media involvement *133*, 133–35, *134*
Merlin (*Falco columbarius*) 5, 24
Middleton, Nigel 133
migration 5–6, 9, 19, 51–52, 135–39, *136*, *137*, *138*
Mistle Thrush (*Turdus viscivorus*) 85

Monash University 186
Montagu's Harrier (*Circus pygargus*) 31
Moorhen (*Gallinula chloropus*) 91, 94, 97, *98*, 98
Moss, Stephen 182
Moult and Ageing of European Passerines (Jenni and Winkler) 108
moulting 24–26, *25, 26*
museums 112, 149, 183
mutes 43–44, *109*
myths 140–44

Natural Childhood (Moss) 182
neck 15, *16*
neonicotinoids *151*, 151
nest boxes 39, 40, *115*, 115–17, *116, 117, 118*
nest making 55, *55*
nictitating membrane 18
Nightingale (*Luscinia megarhynchos*) 127
nightjars 86, 127
North America 9, 36–37, 68, 98, 108, *136*, 139
  see also Canada; United States (USA)
Northern Goshawk (*Accipiter gentilis*) 1, 11, 76, 142, 165
Northern Ireland 138
Northern Wheatear (*Oenanthe oenanthe*) 86
Norwich, England 61–62, 69, *118, 122, 166*, 166
nostrils *11*, 11

Olsen, Jerry 84
orcas 150–51
organisations 191–92
Osprey (*Pandion haliaetus*) 1, 9, 164
ovaries 55–56
overview of Peregrines 5–8
owls *40*, 40, 46, 76, 89, *89*, 102, 166

Packham, Chris 183
parakeets 90
parrots 84, 90
Passive Integrated Transponders 132
pattern of urbanisation 38–39, 49
The Peregrine (Baker) 162
The Peregrine Falcon (Ratcliffe) 54, 68, 163, 187
Peregrine Falcons (Dixon and Shawyer) 116
Peregrine watches 120–22, *121, 122*, 164, *166*, 166, 182, 183, *184*
persecution 70, 145–46, 148–151, 167–68, *173*, 173–74, 179–182, *180*, 188
Persistent Organic Pollutants (POPs) 150
pesticides 36, 149–150, *151*
petrels 19, 88
photoreceptors 17
Pied Wagtail (*Motacilla alba yarrellii*) 75
pigeon fanciers 169, *169*, 170, 174, 177, 178
pigeons
  control of *144*, 144
  gull attacks 37–38
  nearby *78*, 78
  as prey 44, 76–77, *77*, 81, 82–83, *140*, 140–41, 168, *171*, 179
  remains *104*, 105, 111–12
  urban 82, *169*, 172, 172
  see also racing pigeons

plovers 52, 75, 87, *87*, 94, 100, 160
Poland 3, 95, *127*, 146
pollution 150
polybrominated diphenyl ethers (PBDEs) 151
polychlorinated biphenyls (PCBs) 151
polygyny 72–73
popularity 1, 133, 153–54, 162–66, *163*
population densities 36
population sizes 49, 145, 146, *146*, *187*
Powell, Adele 99
power lines 160, *160*
predator-prey dynamics 157–59
predators *40*, 40, 76, 102–3
preening 26–28, *27, 28, 62*
prey 75–76, *77*, 81–82
  bats *92*, 92–93
  caches 44, 52–53, *53*, 79, *100*, 100–101
  caged / escaped birds 90, *90*
  corvids *89*, 89
  gulls 88, *88*
  hirundines and swifts 92
  pigeons *82*, 82–83
  predator-prey dynamics 157–59
  racing pigeons 168–172, *171*
  raptors 89–90
  ringed birds *93*, 93–94
  river birds / waterbirds 90–91, *91*
  seabirds 87–88
  Starlings *83*, 83–84, *84*
  wading birds 86, 86–87, *87*
  woodland, garden, farmland birds 7, *84*, *85*, 85–86, *86*
  see also diet studies
protection 147–48, 179–182, 188
Prytherch, Robin 102

Quail (*Coturnix coturnix*) 86, 94

racing pigeons 82, 159, 168–172, *169, 171*, 174–79, *176, 178*
rails 91, 94, 97, 98, *99*
Raptor Alliance 177, 178
Raptors (Hardey et al) 128
Ratcliffe, Derek 54, 68, 70, 150, 163, 187
Raven (*Corvus corax*) 88, *102*, 102–3, *143*, 143
The Really Wild Show (BBC) 182–83
Red-backed Shrike (*Lanius collurio*) 86
Red-breasted Goose (*Branta ruficollis*) 101
Red-footed Falcon (*Falco vespertinus*) 5
Red Kite (*Milvus milvus*) 149, *154*, 154–55
Red-necked Grebe (*Podiceps grisegena*) 91, 94
Redig, Pat 73–74
redpolls 86
Redwing (*Turdus iliacus*) 75, 85–86, *86*, 95, *95*, 112
research needed 189
Ring-necked Parakeet (*Psittacula krameri*) 90
Ringed Plover (*Charadrius hiaticula*) 52, 160
Ringers' Manual (BTO) 128
ringing / tagging 1, 125–28, *129–130*, 188, *188*
  checklist 132–33
  colour rings 50, *50*, 51, 127–28, *128*, *130*
  data gleaned 135–39, *136, 137, 138*

licence required 128
media involvement *133*, 133–35, *134*
tips 128
tracking devices 126–27, *127*
risk assessments 120, 121, 123, 124
Roberts, Graham 46, 49, 94, 117
Robin (*Erithacus rubecula*) 78, 79–80
Rock, Peter 37
Rook (*Corvus frugilegus*) 89
Roseate Tern (*Sterna dougallii*) 87–88, *93*, 93–94
Royal Pigeon Racing Association (RPRA) 170, 176
Royal Society for the Protection of Birds (RSPB) 155, 166, *173*, 174, 180, 181
Ruddy Duck (*Oxyura jamaicensis*) 91
Ruff (*Philomachus pugnax*) 87
rural Peregrines 49, 53, 54, 70, 147, 148, 161, 180–81, *181*
Russia 101, 137, 149

safety 124
Saker Falcon (*Falco cherrug*) 5, 8, 9, 39, 160
sandpipers 87
Sandwich Tern (*Sterna sandvicensis*) 87
satellite birds 49
satellite tracking 36, 100, 126, *127*, 127, 135, 137, 139, 188
schools 184–85
Scotland 32, 87, 132, 136, 138, 148, 158, 165
seasons *see* year in the life
Second World War *149*, 149
sexual dimorphism *23*, 23–24, *24*
Shawyer, Colin 116
shearwaters 19, 88
Shoveler (*Anas clypeata*) 91
shrikes 86, *113*
sight *14*, 14–18, *16*, 95
signs of presence 43–44
Silver Gull (*Larus novaehollandiae*) 84
Siskin (*Carduelis spinus*) 86
size 22–24, *23*, *24*, 41–42
skin colour 11, 28, *30*, 30–32, *31*
skull *12*, 13–14
Skylark (*Alauda arvensis*) 95
Slaty-breasted Rail (*Gallirallus striatus*) 97
Slaty-legged Crake (*Rallina eurizonoides*) 97
smell 19
Smith, Ashley 153–54
Smith, Hamish 153
Smith, Judith 103
snipes 87, *94*, 94–95, 97, 98
Sokolov, Aleksandr 137
Sokolov, Vasiliy 137
Song Thrush (*Turdus philomelos*) 53, 85, 112
sounds 43
South American Painted-snipe (*Nycticryphes semicollaris*) 97
Southampton, England 161
Southern Cross Peregrine Project, USA 139
Sparrowhawk (*Accipiter nisus*) 7, *21*, 165
decline 168
eyes 11, *141*, 141
killing and plucking prey 6–7, 109, 141
and pigeons 170, 178
as prey 89
size 22, 142
sparrows 84, 85, 167
speed 19–22, *20*, *79*, 80
Sperring, Chris 102
Spot-flanked Gallinule (*Gallinula melanops*) 97
Spotted Crake (*Porzana porzana*) 94, 95
spotting Peregrines *41*, 41–46
Springwatch 68, *117*, 133–34, *134*
stakeholders 123–24, 175–77
Starling (*Sturnus vulgaris*) 17–18, *83*, 83–84, *84*, 107, 167
The State of Nature 33–34
Stock Dove (*Columba oenas*) 82–83
Stonechat (*Saxicola rubicola*) 86
stoop dives 6, 15–16, 19–21, *20*, *79*, 80, 142
Storm Petrel (*Hydrobates pelagicus*) 88
studying Peregrines 104
diet *see* diet studies
Internet and web cameras *117*, 117–120, *118*, 119
liaising with landowners/stakeholders 123–24
nest boxes *115*, 115–17, *116*, 117
Peregrine feathers *114*, 114–15
Peregrine watches 120–22, *121*, 122
risk assessments 124
subspecies 7–8
success factors 8–9, 34–35
survival rate 32, 79–80
Sussex, England 46, 54, 94, 117–18
Sutton, Luke 107
swallows 72, 92
Sweden 3, 5, 18, 32, 51, 72, 136, 150, 153
swifts 19, *65*, 80, *92*, 92
Symonds Yat Rock, Gloucestershire, England 164

Taiwan 97–98
talons 6, 28, *29*, 76
Taranto, Paolo 38
taste 18–19
taxidermy 149
Teal (*Anas crecca*) 90–91, *91*, 95, *106*, *110*
television programmes 68, *117*, 133–34, *134*
terns 17, *34*, 34, *87*, 87–88, *93*, 93–94
territorial disputes 8, 53, *54*
testes 55–56
testosterone 30–31
Thomas, Mark 180
threats
changes in landscape 156, 189–190
chemicals 149–151, 188
decline in prey 157–59
disturbances 161
hybrids 160–61
injuries 152–53, *153*
wind turbines and power lines *159*, 159–160, *160*
see also persecution
thrushes
as prey 53, 75, 78, *84*, 85–86, *86*, *95*, 95
remains 106–7, 112
survival 79–80, *157*, 157

tiercels 23
tits 17–18, 72, 85
tomial tooth *6*, 6, 13, 80, *80*
tongue 14, 18–19
Tordoff, Bud 73–74
tracking devices 126–27, *127*
Tracks and Sign of the Birds of Britain and Europe (Brown et al) 108
Treecreeper (*Certhia familiaris*) 85
Treleavan, Dick 77
Tribe, Alice 180
Tucker, Vance 15–16, 21
Tufted Duck (*Aythya fuligula*) 91
Tully, John 75–76, 93, 172
Turkey Vultures (*Cathartes aura*) 19
Turtle Dove (*Streptopelia turtur*) 82–83, 95, 127

ultraviolet light detection *17*, 17–18
United Kingdom (UK)
  distribution 4
  identification resources 108
  laws 147–48, 181–82
  persecution 148–151, 181
  population 3, 40
  reporting crimes 47
  ringing / tagging 127, 128, 132
  ringing studies *138*, 138–39
  risk assessments 124
  The State of Nature report 33–34
  subspecies 7–8
  year in the life *see* year in the life
  see also specific sites
United States (USA) 20, 32, 36–37, 73–74, 97, 108, 139, 148
University of Bristol 13, 44, 93, 167
urban environments 33, *33*, 34, *35*, 35, 37, *175*, *189*
urban sites, use of 2, 10, 156
  churches and cathedrals *9*, *23*, *27*, *28*, *29*, *39*, 39–40, *41*, *192*
  industrial buildings *42*, *66*
  office buildings *8*, *43*
urban success factors 34–35

Virginia Rail (*Rallus limicola*) 98
vision *see* sight
vultures 19

wading birds *86*, 86–87, *87*
wagtails 75
Waldrop, Stewart 176–77
Wales 3, 87, 100, 159, 171
Walsh, Peter 53
warblers 85, 97, 158
Water Rail (*Rallus aquaticus*) 94, 98, 99, *99*
waterproofing 27–28
waxwings 86
web cameras
  identifying individuals 50
  monitoring nests 57, 58, 61, *119*, 119
  public interest *117*, 117–18, *118*, 133, 166, 183, 188–89, *189*
  things to consider 119–120
  web addresses 190–91

Wegner, Peter 69
Welsh Raptor Group 3
West Hatch Wildlife Centre 152, 153
wheatears 86
Whimbrel (*Numenius phaeopus*) *87*, 87, *106*
White-breasted Waterhen (*Amaurornis phoenicurus*) 97
White-tufted Grebe (*Rollandia rolland*) 97
Wick Quarry, England 68
Wigeon (*Anas penelope*) 91
wild boar *34*, 34
Williams, Iolo *134*
Wilson's Snipe (*Gallinago delicata*) 98
wind turbines 15, *159*, 159–160
wingspans 24
Wood Warbler (*Phylloscopus sibilatrix*) 158
Woodcock (*Scolopax rusticola*) 53, 87, 94, 97, *98*, 99, *100*, *106*, 126
woodpeckers 85, *85*
Woodpigeon (*Columba palumbus*) 82–83, *83*
Worcester, England 39, 50, *97*
word cloud *163*, 163–64
World Feather Atlas (German Feather Research Group) 108

year in the life *48*
  late summer 49
  autumn and winter 49–53, *52*
  spring and early summer
    nest making 55, *55*
    mating 55–56, *56*
    egg laying *54*, 54–55, 56–57
    incubation 57–58, *58*
    hatching 58–59, *59*
    brooding and feeding 59–61, *60*, *61*
    development, fledging, dispersal 62, 62–63, *63*, *64*, *65*, 65–66, *66*
Yellow-breasted Crake (*Poliolimnas flaviventer*) 97

Zebra Finch (*Taeniopygia guttata*) 17–18